Different but Equal

David D. Weisher, M.D.

Diplomate Neurology

**Faculty Member   Clinical Instructor
Georgetown Univ. School of Medicine**

VANTAGE PRESS
New York

FIRST EDITION

All rights reserved, including the right of
reproduction in whole or in part in any form.

Copyright © 1994 by David D. Weisher, M.D.

Published by Vantage Press, Inc.
516 West 34th Street, New York, New York 10001

Manufactured in the United States of America
ISBN: 0-533-10519-6

Library of Congress Catalog Card No.: 92-94298

0 9 8 7 6 5 4 3 2 1

To Mr. Charles Rathbun, the man who changed my life

Contents

Foreword	ix
Preface: Cerebral Lateralization and Understanding Ourselves	xi
Acknowledgments	xix
My Story	xxi
1. The Concept of Cerebral Lateralization	1
2. Sex and Lateralization	25
3. Homosexuality	50
4. The Concept of Soul/Individual	66
5. Savant Syndromes and Genius	84
6. Sex, Behavior, and Lateralization	106
7. Memory	123
8. Conclusion	147
Bibliography	161

Foreword

This book is a layman's guide to cerebral lateralization and brain mechanisms. It illustrates how lateralization is involved in sexual differences, memory, sexual preference, and behavior. The three main axioms of lateralization theory are as follows:

1. The brain is divided between left and right hemisphere skills.
2. Many of these skills were determined in utero (prior to birth).
3. Behavior is influenced by these skills.

This theory is gaining strength almost daily and is in direct contrast to Freudian theory. New scientific data supports lateralization and negates Freudian theory, which has a paucity of objective scientific support.

The book is fun, entertaining, supports itself well scientifically, and is based on seven years of literary research.

The author believes that there is too much unauthoritative work in this field for public consumption. Much is written on the left versus the right brain; however, little is correct or useful. The reason why John does not ask for directions is not because of ego but because he sees his universe from the right hemisphere (visual-spatial) and does not feel comfortable going to his less functioning, left hemisphere (verbal-temporal), asking directions and then converting back to the right. You see, it has nothing to do with John's emotional gratification, but everything to do with how the Good Lord hardwired his brain. By understanding our brains, we grow to love and understand each other.

Included with the book is an Ocular Dominance Wall Chart. By following the simple instructions, the reader can discover his/her dominant eye (left or right) and how strong it is. Then the reader can determine how he/she fits into the behavior patterns in the book. Are the eyes the "window of the soul"?

All in good fun. . . .

The author has an extensive and varied background that enables a unique approach to brain science.

He was, at one time, an aerospace computer engineer working on the flight computers for the F-14 Tomcat, the B-1 bomber, and even a little on the space shuttle. He was also a deep-sea diver, trained in mixed gas, saturation, and bell diving; he is now a medical doctor and Board-certified neurologist, with a fellowship in electrodiagnostics from Georgetown University. He is also on the faculty of Georgetown University School of Medicine, Department of Neurology, as a clinical instructor.

The author practices neurology in Greenbelt, Maryland.

Greenway Neurology Ass.
7525 Greenway Center Drive
T-8
Greenbelt, Maryland
301-982-4552

However, Dr. Weisher feels that his most important achievement is surviving the effects of being dyslexic and growing up in the sixties. As a youth he underwent extensive psychological testing and was told that he was mentally retarded and not fit for college. It was recommended that he become a barber. However, in high school he was found to be skilled in advanced math and the principal, Mr. Rathbun, convinced him to go to college and become an engineer. (This is the reason for the dedication of the book.) By understanding cerebral lateralization, the author discovered that his condition was a point on the spectrum of lateralization and thus not inferior. He was "different but equal."

—Charlotte B. McCutchen, M.D.
Medical Office and Health Science Administrator
National Institutes of Health, Bethesda, Md.
Associate Professor of Neurology
Georgetown University School of Medicine
Washington, D.C.

Preface: Cerebral Lateralization and Understanding Ourselves

With advancements in medical technology and research, there has been a revolution of thought regarding brain function and its relationship to behavior. Previously, it was felt that environment had an exclusive influence on behavior. Today we are learning that behavior may follow skill and that many skills were provided in utero (prior to birth). Therefore, the characteristics that make us unique are likely to be predetermined prior to birth.

Freud and Cerebral Lateralization

Freud, the "father" of modern psychiatry, recognized differences in male and female behavior patterns and attempted to explain them. With the paucity of scientific data at the time, he assumed that environmental influences played a significant role in behavior. Freud emphasized experiences occurring early in life as determinants of adult behavior, the first three years of life being of particular importance. Both environmental and congenital influences are important in the ultimate behavior of the individual. However, what characteristics are influenced by environment and what characteristics are solely influenced by congenital causes? The battle between environment and heredity is far from over. There's an ocean of data on environmental influences and some is not very clear. However, there is now scientific evidence suggesting a congenital influence on later behavior.

Cerebral Lateralization and Behavior

Cerebral lateralization is more than just the study of skills unique to the left or right hemispheres of the brain. It is a revolutionary idea that our brains are machinelike in function, divided into various skill areas. These skills were developed prior to birth and may later influence behavior. Cerebral lateralization might explain why we do "crazy" things, why we select certain hobbies, fall in love, have sexual attractions, become homosexual, or remember, if you are a woman, things that men forget.

The concept that our brain might be divided into areas of different skills is not new. Dr. Karl Kleist published a book in 1934 on neuropathological observations for approximately three hundred brain-injured soldiers from World War I. He presented a "brainmap;" showing various areas of the brain that he felt were responsible for psychic functions. However, his book was severely criticized and did not gain wide acceptance. Cerebral lateralization carries this one step further by localizing function, not only to specific brain areas, but into two major groups: the left and the right. The left hemisphere skills are verbal/temporal and the right, visual/spatial, equally representing the verbal and nonverbal world in our brains.

Sex and Lateralization

If there are two different sexes and two different hemispheres in the brain, is it possible that one hemisphere may predominate in a particular sex? Is there scientific evidence to suggest this?

Scientists have known for a long time that we can learn about ourselves through animals. In 1981, Diamond, Dowling, and Johnson reported an interesting observation in rat brains. They measured specific areas in the cortex (outer shell) for thickness. They found a statistical difference between corresponding areas when male and female rat brains were compared. The females had thicker left hemispheres while the males had thicker right hemispheres.

In 1969, Taylor studied children with febrile seizures who

later developed temporal lobe epilepsy. He reported two interesting conclusions: (1) the left hemisphere matures after the right, and (2) the male brain matures after the female. These studies revealed evidence of a difference in function between the hemispheres as well as differences between the sexes. Educators have observed that girls excel during the early years in school. This might be explained by left hemisphere orientation (reading, writing, and arithmetic) in early education, which gives the female brain a gender advantage. The higher incidence of hyperactivity and developmental delay in boys as compared to girls further substantiates a difference in brain development and function.

Common Examples of Sexual Differences

I have observed for many years how the sexes differ when taking directions. Women more frequently write them down (long hand), while men will often draw a little map. Cerebral lateralization clearly explains this. The female is simply mobilizing her cerebral energies to the naturally higher functioning left hemisphere (verbal/temporal). We use the skills we have. The male, on the other hand, mobilizes his right hemisphere (visual/spatial) by drawing a map. Both work equally well. They are functionally equivalent (different but equal).

Few will argue that sex and romance are the common denominators of human behavior. Today, well over 5 billion people testify to the success of marketing to arouse sexual interest. However, the means of arousing that interest is often different depending on the gender of the focal group. If you don't believe this, just look at what sells in the periodical section at the supermarket or bookstore. The material marketed for women facilitates the verbal/temporal left hemisphere, romance novels, etc. The male mind, on the other hand, is marketed through the right hemisphere (visual/spatial) with the "centerfold." The object, "sex," is the same; only the means of stimulating interest is different. Again we have evidence of behavior following skill. Notice how *he* is always "whispering sweet nothings in *her* ear."

Cerebral lateralization supports cognitive differences. We

are now learning that this concept is far more than just skills unique to each hemisphere. It may explain genius, homosexuality, heterosexuality, memory, as well as our selection of a career.

Thus, one might conclude that if the world revolves around anything, it revolves around the concept of cerebral lateralization, for this concept is the framework of all that we see in the world. We see different people with different skills and desires doing and accomplishing different things. We observe men and women, often quite different from each other, yet attracted to one another. Why should the difference be so attractive? We learn from the concept of lateralization that nature prefers variety and difference, yet we often buck the system in our own lives by stressing sameness. "I don't like her; she's different" or "Don't associate with him; he's weird, if you know what I mean." Cerebral lateralization tells us that we are all unique yet equal in our ability to function and resolve problems confronted in daily living.

The Development of Lateralization Theory

Prior to World War II, neurologists believed that there were dominant and nondominant hemispheres in the brain. One side of the brain was the master, the other the slave. Right-handed people were believed to be left hemisphere dominant and left-handers, right hemisphere dominant (dominant meaning that side of the brain responsible for the enunciation of speech and the interpretation of language). Therefore, loss of the dominant hemisphere would result in the inability to talk or understand language and communicate or interact with others. Many believed that environment played an important role in the development of cerebral dominance. If a child was encouraged to use his left hand, then his right hemisphere would become dominant.

After World War II, the follow up of brain-injured veterans revealed that most left-handers with right hemisphere damage continued to effectively communicate. This was clearly not consistent with the old theory of a dominant and nondominant hemisphere. It was discovered that each hemisphere, left and right, was endowed with unique skills, the left being verbal/temporal and the right, visual/spatial. There was no master/slave relationship between the hemispheres, rather a harmonious working relationship between the verbal (left) and nonverbal (right), a sort of different but equal relationship. Neurologists continue to use the terms *dominant* and *nondominant,* however, not in the same context.

If each hemisphere is endowed with unique skills, then environment may play a less important role in the overall development of personality than has been assumed. We are learning that a significant part of behavior is determined prior to birth. Our grandmother may have been right when she said, "The acorn never falls far from the tree," or "Like father like son." In this book, I show evidence that handedness (left or right) is determined prior to birth. One might be able, by examining an infant of just a few weeks, to determine with 95 percent accuracy the future handedness of the child. Perhaps the "hardware" is just as important, if not more so, than the "software."

President Bush has called this the decade of the brain. I feel that this book is, therefore, appropriate to be printed at this time. For we are our brains. We are not our liver, as the Babylonians thought, or our heart, as the Hebrews believed. Rather it is the brain that is the origin of thought. It is the brain that is the seat of the individual. We are all unique and composed of an aggregate of various left and right hemisphere skills that influence our future careers, hobbies, and personalities.

Cerebral lateralization is more than just the study of various functions unique to the left or right hemispheres of the brain. It provides evidence that our brains are machinelike in function, with behavior following skills that are predetermined prior to birth. Each of us is endowed with various skills that influence our personality and behavior. We see evidence of this machinelike mechanism in nature, ecology, evolution, etc. Why can't the brain be mechanized as well? This does not detract from the concept of God the creator but rather enhances it. King David said in the Psalms that "we are fearfully and wonderfully made." Cerebral lateralization looks at the end product and marvels at the divine design. This book explains the concept of "man the machine" and lateralization and how the latter influences our lives. Sex, homosexuality, genius, behavior, development, memory, and learning are discussed in relation to cerebral lateralization.

Neurophilosophy

The purpose of this book is to provide the lay person with an opportunity to appreciate and enjoy the intellectual recreation that this subject affords. I recall many conversations with my colleagues at parties where we frequently became philosophical, discussing this new concept of cerebral lateralization and how it fits into the scheme of things. If the brain is divided into various skills (left and right), are we just a product of some divine engineering? Are we just machines (androids), or is there something more that we have yet to learn? We believed that only a neurologist could really discuss these ideas. However, I feel that anyone who is curious would also have fun with this fascinating concept. It is for this reason that I have written a book. This book is my way of introducing cerebral lateralization to the general public. It is intended to provoke interest and develop an appreciation of our brain and thus life itself. It is not without some novel ideas of my own.

The study of cerebral lateralization might assist us in attempting to understand many perplexing characteristics of the human brain. For example:

Why are there special boy and girl toys?
Why are most homosexuals male?
Why do many women buy romantic novels?
Why are most idiot savants male?
Why are women and men so different?
Is homosexuality predetermined prior to birth?
Why do men like to look at nude girls?
Where is the origin of thought (self)?
How much behavior is determined prior to birth?
What is the cause of dyslexia?
Why is there more hyperactivity in boys?
Why do men forget while women often remember?
What is inspiration and intuition?
What is the "late bloomer"?

The first chapter acquaints the reader with basic brain mechanisms as a background for understanding the other chapters. It also introduces the concept of cerebral lateralization.

The second chapter is an armchair discussion of the differences between male and female brains. Why is a difference between the sexes likely? There are two sexes and two hemispheres. Is it possible for a hemisphere to predominate in a specific sex? If there is a difference, how can it be described? Are the brains of the male and female different although functionally equivalent?

The third chapter looks at homosexuality from the point of view of cerebral lateralization. For years, experts have suggested that homosexual behavior is the result of environmental influences (Freud). However, with current evidence, it appears that many male homosexuals achieved their sexual preference in utero (prior to birth) and thus had little "choice" in their desires.

The fourth chapter discusses the concept of the individual and soul. Cerebral lateralization views the brain as a machine with different sites representing different skills. If this is true, then where is the "soul," that part of the brain that is the individual (can no longer be divided)? No clear cut answers are presented, only "food for thought."

The fifth chapter explains savants and genius. When viewed from the perspective of lateralization, there is a better understanding of these unusual skills. The calendar brain, musical talents, and artistic abilities are discussed from the conceptual viewpoint of cerebral lateralization.

The sixth chapter brings together the myriad of personalities and behavior patterns that may be the result of cerebral lateralization. Lateralization explains congenital influences in our choice of hobbies and career selection. This chapter explains how nature (through lateralization) favors diversity. Perhaps communism, with its central planning, failed because it goes against the natural process.

The seventh chapter looks at memory and learning in relation to cerebral lateralization. Memory and recall are specialized functions using various areas of the brain. Near-death visions may, in fact, be part of this same mechanism. Man, the religious animal, is discussed, and the possibility that religious information might actually be processed differently.

The final, eighth, chapter discusses why cerebral lateralization influences relationships. Perhaps with some knowledge of cerebral lateralization we will develop a better understanding of each other. Behavior patterns typical of the right brain person and the left brain person are also presented.

The book comes with an Ocular Dominance Chart that readers can use to see if they are right or left eye dominant and thereby gain some insight about their own brains. There are no conclusive studies showing a relationship between ocular dominance and left or right hemisphere cognitive skills. It is hopeful that this book will stimulate further research.

All in good fun. . . . Perhaps if we know about ourselves, we can learn to live with others, for cerebral lateralization teaches tolerance, that we are all different but equal and just as important a part of nature as anyone else.

Acknowledgments

Thanks to Charlotte McCutchen, M.D., for her advice, support, and encouragement.
Thanks to Sheryl Burg for her editorial assistance.
Thanks to Desmond O. Doherty, M.D., for his encouragement.
Thanks to Warren Rasmussen, M.D., for artistic advice.
Thanks to Aleem Iqbal, M.D., for his encouragement.
Thanks to my Georgetown neurologist friends for our many discussions at parties, which contributed to this book.
Thanks to Judith Feldman for her editorial assistance.

My Story

I was born in Boston, Massachusetts, in 1950 to good and intelligent parents. I must admit that my life was not hard in the sense of having material things, for my father and mother were good providers. They were also good providers in a spiritual and intellectual way as well. I was a very curious child, taking apart everything I could find. I remember totally dismantling my father's Bulova watch at the age of six with a jeweler's tool set. My father must have been able to relate to my greed for knowledge for he neither scolded nor punished me. He simply put the parts in a bag and presented it to the jeweler, who reassembled the watch for him. I had a remarkably understanding father. I was always asking questions. What made cars go? Where do babies come from? There was no end to my thirst for answers.

I started school in Hartford, Connecticut, at the age of six years. I desperately wanted to learn to read as I realized its value for gaining knowledge. However, in school I was hyperactive and was unable to learn anything. I recall not wanting to return to class from recess. It did not take long for parents and teacher to agree that either I was not ready for school or school was not ready for me. I was then sent home to try again the next year.

At the age of seven years I was readmitted to the first grade at the same private school. Although less hyperactive, I was clearly not learning anything. The teacher felt that this might be because of my eyesight, and after seeing a doctor, I was fitted for glasses. I continued to learn little to nothing. Parents and teachers were perplexed for I had come from a fairly bright family. I was simply passed year to year. My reading skills were especially poor, and even though in the fourth grade, I was still doing reading exercises with the second graders. You might say that I was an intellectual "basket case." I was passed on year to year until it was decided that the seventh grade would be done in a public school, Leland P. Wilson Junior High in Windsor, Connecticut. You must keep in mind that this was in the middle of President Johnson's "Great Society" of the 1960s. Uncle Sam was going to show and tell us all the way.

I should point out that although my interest in schoolwork was at an all time low, my interest in science and chemistry was very high. I could account for each stage of the carbon cycle, which is the nuclear reaction of the sun, and I was the only kid in town with a functioning cloud chamber (nuclear physics lab for observing alpha particles). However, school did not interest me; therefore, my mind was closed to it.

At Leland P. Wilson Junior High School I continued to fulfill my mission of mediocrity. Although knowledgeable in science, my grades in science did not reflect this for my spelling was awful. My teacher, being intelligent as well as observant, felt I might need special help and evaluation. My parents agreed and psychological evaluations were started. I was even put into special education classes. Special education was unforgettable and perhaps unforgivable. My fellow classmates were clearly and significantly mentally retarded, often unintelligible. It was here that I started to believe that I was truly inferior. My original classmates were starting to think of college. For me it was clear that college was out of the question. College was not for the inferior; college was for the normal.

It was obvious that I had no real interest in school. I recall that my English studies included the book *Johnnie Tremain,* a fictional story of a young boy growing up in revolutionary America. Although history has always been an interest of mine, my reading skills were too poor, and probably because of that fictional material was of no priority. My grades were mainly Ds and Fs; a C would have been cause for celebration. To be honest, I'm not sure I got a C in the entire seventh grade.

My evaluations by psychologists continued that year until spring when they presented their final impression to my mother. The presentation was given over the phone, and I listened on the other line. They concluded that I was definitely not college material and recommended trade school. I recall a suggestion that because of my independent ways and desire to work with my hands, I should probably be a barber. To be honest, I took this to heart and started to lay plans to go into the hair-cutting business. It is a fact that from that time to this day I cut my own hair. I will also admit a certain affinity for those in this noble and honest profession. Years ago, the country barber was considered similar to a doctor. However, I'm sure the psychologist did not have that in mind at the time of our meeting. Thus, you might conclude that I was labeled inferior, and a very real intellectual barrier was put in place.

I began private high school that fall with no plans of ever going to college. My new principal was notified of my relatively poor intellectual potential but accepted me anyway. I must admit that there was one special thing that troubled me. If I was so dumb, then why was I so curious? I continued to read about science, navigation, technology, electronics, etc. I loved music, and having no formal music education I was able to create all kinds of pieces on the piano. At that time, I had no explanation for this skill. However, being told by experts that I was inferior, I was not about to "rock the boat."

It was in this private high school that my talents were discovered. To be more specific, it was the study of math (geometry) that started the principal, Mr. Rathbun, thinking differently of me. He had been told that I was inferior but to let me pass because my parents donated freely to the school. However, if I were so inferior, how was it that I was getting 100 percent on my math scores?

Mr. Rathbun thought this was so interesting that he invited himself over to my parents' house for dinner one evening and insisted he would not leave until I promised I would go to college and study engineering; he would help with the arrangements.

I wrestled with this thought a great deal for I had always thought of myself as inferior. This is what the "experts" had told me. Now an "expert" was saying something very different. After discussing this with my parents, I agreed to go to college (engineering) and become a computer expert. I discovered that I had a fairly creative mind as well as a natural talent with digital logic and Boolean algebra (the mathematics of computer hardware), right hemisphere skills typical of dyslexics.

It was the study of computers that triggered my interest in the human brain and the desire to know more about myself and people like me. For that reason, I gave up aerospace computers and entered medicine. I then had the opportunity of studying neurology at Georgetown University, which afforded me the privilege of serving the medical-neurological needs of some prominent and famous people. However, just prior to graduating from medical school, during a time when my mother was near death from Lou Gehrig's disease, we drove to New England and found this wonderful teacher, Mr. Rathbun, and personally thanked him for changing my life. Thus I have only the highest regard and respect for the responsibilities of the teaching profession.

Summary

The reason for telling this story is that the human brain is the most complex organism known to mankind. There is no way that we can summate all of its skills in a simple examination, be it three hours or three days, for there may be a complex myriad of various left or right hemisphere skills in verbal-temporal or visual-spatial subjects. Little was known about dyslexia then. Less was known about the possibility of right hemisphere skills. Don't be discouraged about the young child who appears a little slow to learn. He or she may later show focal talents, "islets of genius," and prove the "experts" wrong. We should always be very slow to label a young brain inferior. By doing so we may rob the world of a valuable talent, for who can measure skills such as insight, creativity, or even "subconscious logic"? When you stop and think, you may realize that these skills are the framework of our worldly culture. When I listen to Mozart, I appreciate the skill of the musicians, but who can doubt the genius of the composer? For they will always call it Mozart and nothing else. Perhaps there is no such thing as a "late bloomer," just undiscovered talents. Always encourage the young brain. After all, it was designed by God. Let no man put it down.

The study of cerebral lateralization, as well as neurology, was of great importance to understanding myself. I learned that I was extremely "right hemispheric" in cerebral function, which had an effect on my behavior as well as on my abilities. I learned that I, too, was "different but equal." I hope that this study will be of great help to you as well.

This book is dedicated in loving memory to Mr. Charles Rathbun.

Different but Equal

1. The Concept of Cerebral Lateralization

In order to understand the concepts described in this book, a discussion of basic brain mechanisms is helpful. This allows readers from all walks of life to start out from a similar reference point. Included in this chapter is an introduction to the concept of cerebral lateralization.

The Organ We Call the Brain

The human brain is the crown of creation, the final frontier, the ultimate of man's intellectual desires. For locked in the human brain are the secrets of nature. One may feel that to know ourselves is to know God. We are our brain. We are not our liver, which the ancient Babylonians and Sumerians thought was the origin of intellect. Hippocrates (400 B.C.) was perhaps the first to realize that the brain was where thought originated. He also might have been the first to discover that damage to one side of the brain caused paralysis of the other side of the body. For example, destruction to the right hemisphere will cause paralysis of the left face, arm, and leg. This is probably the most fundamental concept of understanding brain function. It was over two thousand years later that Fritsch and Hitzig (1870) helped to shed light on this concept by applying electric charges to one hemisphere of a lightly anaesthetized dog, causing movement of the opposite side of the body.

In 1861, at the Societé d'Anthropologie in Paris, a gentleman by the name of Pierre Paul Broca was to change man's concept of his brain forever. Dr. Broca demonstrated that our brain is not just a simple mesh of neurons, but is compartmentalized into

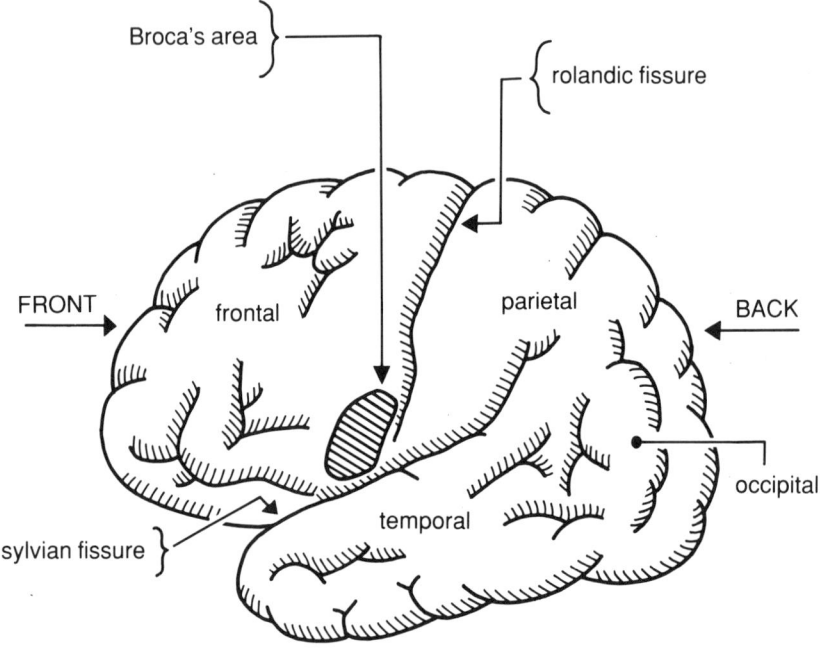

Fig. 1. The four lobes of the brain.

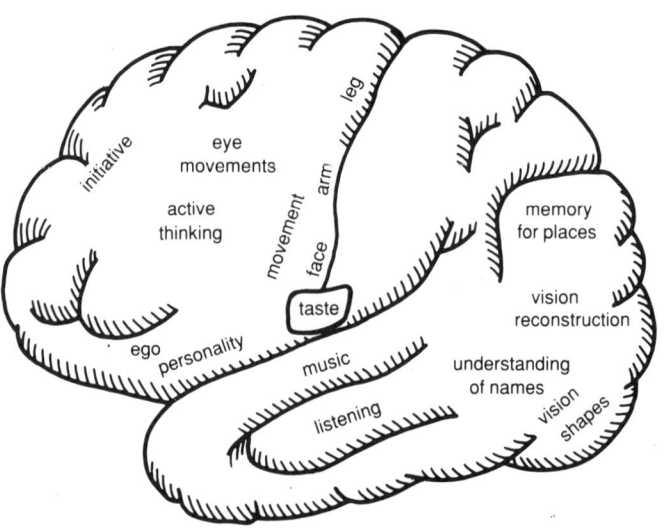

Fig. 2. The brainmap of Karl Klist.

functional areas of neuronal networks. He discovered that a relatively small area of destruction in the dominant hemisphere (which would be the left in those of us who are right-handed) can cause an inability to talk. He was able to prove his point by an autopsy performed on a patient who had this same problem. We call this inability to talk aphasia. This patient was able to perform all other neurological functions, walking, listening, reading, etc. A patient will often describe this as knowing what one wants to say but being unable to get the words out. We now call this condition Broca's aphasia in honor of him. The actual area of destruction was the posterior third of the left third frontal convolution of the brain. This is also called "Broca's convolution."

Many years later, in 1934, Karl Kleist published an interesting book of neuropathologic observations on almost three hundred brain-injured soldiers from World War I. Kleist helped us to further understand this concept of brain compartmentalization; however, he was mainly interested in the psychic functions of the brain and was severely criticized for trying to limit these functions to specific areas of the brain. Then in 1959, Wilder Penfield and Lamar Roberts published *Speech and Brain Mechanisms*. This book was a compilation of the results of microelectric stimulation to areas of the brain during brain surgery while patients were fully awake (about 530 cases). The results of these experiments seemed to indicate that psychic phenomena may indeed be localized to specific areas of the brain, as had been suggested by Kleist. We will discuss this more in another chapter.

In order to appreciate cerebral lateralization and this mechanistic approach to brain function, we will need to discuss some basic neurology. I will now be so bold as to fully explain the basics (or Brain 101). It is really quite simple. The brain is divided into four lobes on each side of the two hemispheres: frontal, parietal, temporal, and occipital. In each of the lobes we find that the brain has various folds, which are called convolutions or gyri. Between the folds are valleys, which are called sulci. When these sulci are large and deep, we call them fissures, such as the rolandic and sylvian fissures.

The Occipital Lobe

The most posterior lobe is the occipital. We see with this lobe. The eye is actually an extension of the brain and connects directly to the lobe. Visual information from the eye first enters the central posterior portion and is assimilated (or made sense of) as the impulses move forward through the brain. As you already know, the left brain functions for the right side of the body and vice versa. The same applies to the occipital lobe. The left lobe assimilates visual information for the right visual fields of *both* eyes and vice versa. That is to say, if the left occipital lobe was not functioning, one would only be able to see through the right visual field of *both* eyes.

The assimilation of visual data is truly amazing, especially in light of the fact that this information is stored three-dimensionally and is constantly being compared to memory. This is how we recognize our mother's face and the face of people we meet. There is an instantaneous pattern recognition subconsciously. Millions of bits of information are processed simultaneously. This differs from a computer in a very important way. Computers process information serially, while the brain is a massive parallel processor. A computer is able to process data a million times a second, while the brain will say, "Here are a billion bits of information. They're available now! Where do you want them?" It is in the functioning of the occipital lobe with its massive visual pattern recognition that one can appreciate this parallel process. The brain, although significantly slower than the computer, is capable of processing information in an amazing volume that would overwhelm any man-made computer quickly. Thus, by the time the visual signal arrives near the parietal/temporal regions from the posterior portions of the occipital lobe, they have been well "processed," as a computer engineer would say. Neurologists refer to this processing of data as "association." That is to say, the pattern is recognized and compared to memory, which can result in a deeply emotional response, such as in the case of the face of a loved one not seen in years. These emotional responses are believed to arrive from deeper structures in the brain called the limbic system. We will get to that later.

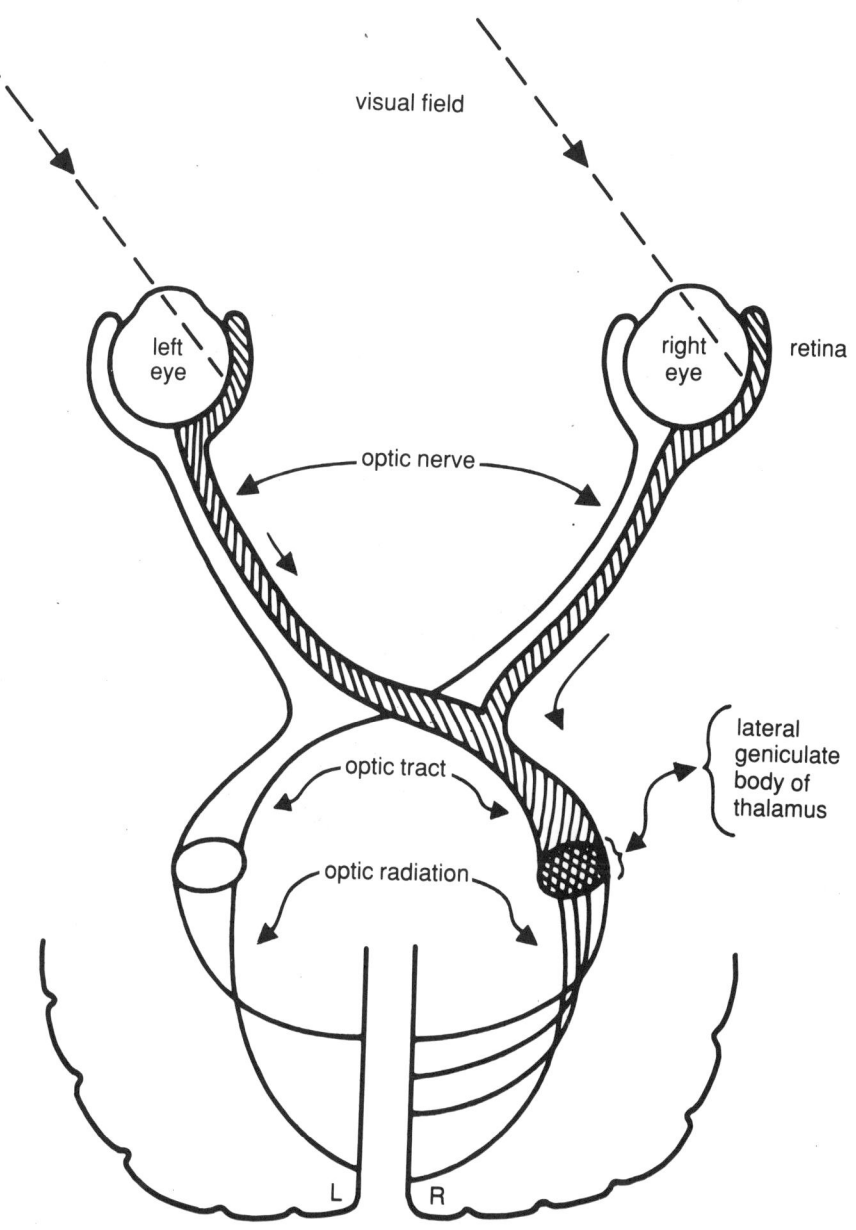

Fig. 3. The occipital lobe: the *left* visual field goes to the *right* occipital; the *right* visual field goes to the *left* occipital.

The Frontal Lobe

The frontal lobe deals much with motor control (muscles) of the rest of the body. While the anterior (front) portions deal with more esoteric forms of motor control (initiative, judgment, decision making), the posterior (back) deals with direct motor control. Starting from top to bottom: legs, arms, face, and mouth.

In the large area between these two regions lie the "frontal eye fields." This area is dedicated to the amazing ability of conjugate eye movements. This is where the initiation of voluntary eye movements take place. Few people really understand the amount of brain matter involved in eye control. Each eye is controlled by six muscles; therefore, all twelve muscles must work harmoniously so that the image falls on the same area on the back of each eye (retina). In the case of a disparity of any one of these muscles the result is diplopia (double vision).

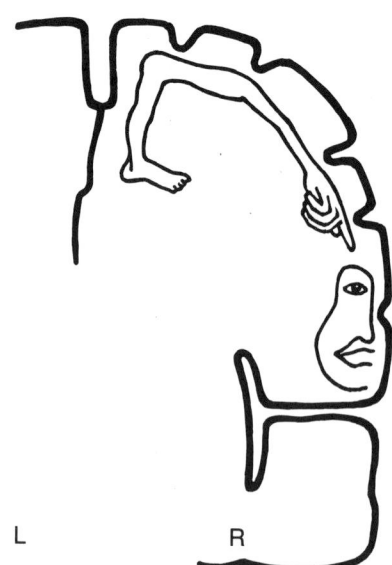

Fig. 4.   The motor areas or the precentral gyrus.

Now moving back to the motor strip or precentral gyrus, we come to our famous Broca's area near the area of mouth control. In those of us who are right-handed, this portion on the left side of the brain would be where speech is initiated. We would therefore call this left side the dominant hemisphere. Many true left-handers will have Broca's area in the right hemisphere. We call this being right hemisphere dominant.

The frontal lobe also deals with behavior. It is considered to be responsible for judgment. I will give you a case story.

I had a very interesting patient named Darleen. She was a most pleasant middle-aged woman, who had been having a profound problem with alcohol. She just wanted to drink gin every day. She was about to seek professional help for this when, one day, she suddenly had a seizure. That's when she came to see me. She looked fine on neurological exam except that she tended to fall to the right when standing with her eyes closed (positive Rhomberg sign). For that reason, a CAT scan of the brain was ordered. This scan (picture of the brain using X rays) showed a large, benign (not malignant) tumor pressing on the right frontal lobe. The patient went to surgery and had the tumor removed. Her follow-up neuro exams have been normal as well as the follow-up EEG (brain wave test). What is interesting is the fact that she stopped drinking after the surgery. She completely lost her desire for liquor after the tumor was removed. This case is important because it suggests the frontal lobe's involvement in judgment, habit, and addiction as well.

The Parietal Lobe

The parietal lobe sits between the frontal and the occipital lobes. This lobe has a "strip" similar to the motor strip of the frontal lobe; it is called the postcentral gyrus, and this is where sensory data is processed or associated. Notice the large cortical area dedicated to the thumb. Think of this the next time you see a woman feeling a fine silk dress. You will observe that she uses her thumb. Now you know why. Notice also the large cortical area dedicated to the lips. This could in part be why we enjoy kissing

Fig. 5. The sensory areas or the postcentral gyrus.

those we are close to rather than rubbing elbows. This is also an excellent example of how behavior follows a skill that is "hard wired" into the brain.

The Temporal Lobe

This part of the brain is perhaps the most fascinating because this lobe has such a varied functional capacity. Often when a nondominant temporal lobe is removed surgically, the patient and friends of the patient do not detect any difference in the patient's behavior or motor skills. The key, of course, is removal of the *non*dominant temporal lobe. A lawyer patient of mine (seen at Georgetown University Hospital) had a severe seizure disorder emanating from his right (nondominant) temporal lobe. Upon removal of this lobe, the seizures ceased, and the patient lived a happy, normal life.

The dominant temporal lobe often has very important func-

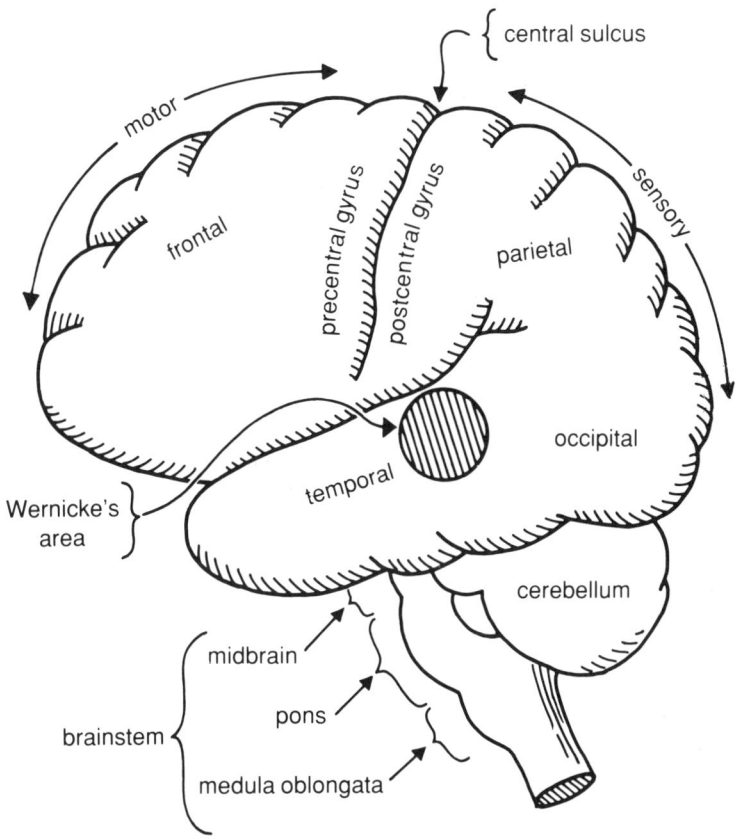

Fig. 6. The temporal lobe and Wernicke's area.

tions. One of these is in the posterior portion of the lobe, which deals with interpretation of auditory language. People with lesions in this area are not able to correctly interpret what people say to them even if they hear the sounds. This syndrome of sensory (receptive) aphasia was discovered by Carl Wernicke (1874) and for that reason it is often called Wernicke's aphasia.

The inner portion of this lobe helps to interpret smell and often, on the dominant side, deals with memory formation. The close proximity of smell and memory has many interesting implications that we will explain in another chapter. I have seen patients who have had strokes involving the destruction of both inner (mesio) temporal lobes, and they are most interesting. You can walk into a room and they will seem fine. They will have no problem with the conversation. However, if you leave the room only to return a few minutes later, they will deny that they have ever seen you. They are simply not capable of forming new memory. This is the result of destruction to an area I call the "data entry zones." This is an area of the brain that information to be stored must pass through. If it is unable to pass, the information is not stored and hence no memory is formed.

The posterior portion of the temporal lobe is also very interesting. Wilder Penfield (1959) discovered that by stimulating this area microelectrically, patients would describe what Penfield called "psychical interpretive" responses. The patients felt they had been there before and believed they knew what would transpire next. This phenomenon is known as déjà vu, which of course means "already seen." Patients with temporal lobe seizure disorder will sometimes describe this phenomena prior to a seizure.

I have another interesting story about the temporal lobe. At our office in Greenbelt I will often read EEGs (brain wave test) on patients I have never seen. One day I read an EEG that showed seizure activity confined to the temporal lobe. I then called the doctor who had referred the patient and an arrangement was made for a consultation. Her name was Donna and she was a young, attractive, and pleasant woman. During the interview she stated that she never had loss of consciousness but described episodes of feeling that she was spinning or moving (true vertigo). These would last for several minutes and sometimes she would feel tired afterward. A diagnosis of "simple partial seizure" was made and the patient was placed on seizure medication. The episodes promptly stopped as long as she took her medicine. In fact, several anticonvulsants were used, and the patient was symptom-free as long as she was on one of them. Thus, we have evidence of the temporal lobes being involved with spatial awareness (where our bodies are in space) as well as time (déjà vu), as previously described.

Summary of the Lobes

The areas of the brain under discussion (the frontal, parietal, occipital, and temporal lobes) are all involved with a very high level of brain function. This would include consciousness, initiation, and interpretation of the environment, as well as intellectual capacity. The very outer layers (there are six) are often referred to as the neocortex (*neo* is Greek for "new"), thus giving credence to the fact that the human cortex is the most phylogenetically advanced of all species.

The two halves or hemispheres of the brain are connected together at three basic areas by insulated (myelinated) nerve fibers. These areas are: anteriorly, the corpus callosum; inferiorly, the anterior commissure; and posteriorly, the splenium.

The Brain Stem

The brain stem is the very central and deepest part of the brain. It is composed of the midbrain, the pons, and the medulla oblongata. To be simple and brief, this area primarily deals with life-support systems (blood pressure, heart rate, sleep, respiration). It is interesting to note that if this area is pressed by the finger, one can induce unconsciousness, bradycardia (slow heart rate), strenuous breathing, and tonic posturing of the body (Foerster and Spielmeyer).

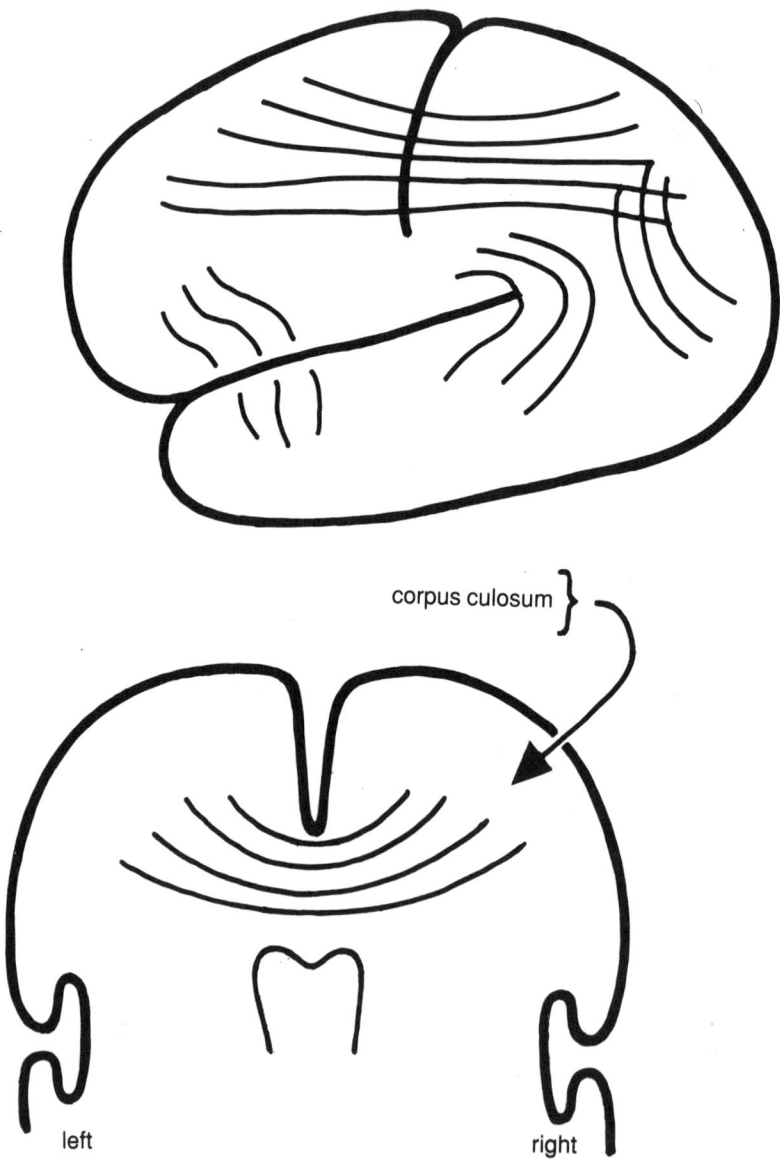

Fig. 7. White matter fibers connecting the lobes.

The Cerebellum

This part of the brain is also called the hindbrain because it sits just behind the brain stem. Its primary function is motor control to ensure that all movements are smooth, a sort of servo mechanism, as an electrical engineer would say. Think of the cerebellum when you see a ballet or a basketball game.

What about Gray and White Matter?

The cortex, brain stem, and cerebellum are referred to as gray matter. This is because of their gray color. Gray matter has a tremendous interlocking of neuronal networking and nerve cell bodies. However, just underneath this mantle of gray matter there is the "white matter." White matter lies between the mantle and the central gray matter (thalamus and brainstem). It is white because of myelinated nerve fibers. We can look upon these fibers as wires (or bus lines) carrying complex information from one part of the brain to another or from one half of the brain to the other. I have believed for many years that in the healthy adult human brain one part can potentially communicate with any other part of the central nervous system.

Where Is "Self" or the "Individual"?

The brain is thus an amazing organic computer, processing data here, then sending it over there for more processing, then committing it to memory and generating an emotional response, etc. All this causes one to think. If your brain is you and your brain is further subdivided into various functional parts, then where is the area that is truly you (the self or individual)? Where is the area that orchestrates all the rest? Who is the "me" that owns all of these things?

Well, you might answer: "We are our *whole* brain," and to an extent you would be right. It is well known that the body's defense mechanisms dictate that the brain is to be protected at all cost.

After all, the brain is enclosed and protected in this cranial vault we call the skull. However, it is also true, as we have seen, that the brain is composed of parts much like the human body. If we lose part of our brain, we do not become another person.

I recall a patient, a casualty of the Iran/Iraq war, who had had half his brain blown off by shrapnel. He was still able to communicate because it was his nondominant side that had been destroyed. Was he any less the "individual" he had been before? It was clear that "self" was still preserved even though a large portion of the brain had been lost. So where is "self"? No one truly knows. However, we will investigate this later in chapter 4.

What Is This Left Brain/Right Brain Thing?

When we talk of the right brain and the left brain as having unique skills, we call this "functional asymmetry." We have also discussed the concept of "cerebral dominance." That is to say, the left cerebral hemisphere dominates communication skills (receptive and expressive) in persons who are right-handed, a concept that came about in the latter part of the nineteenth century. This theory suggests that there is a master/slave relationship between the hemispheres. Thus, if one were right-handed, one's left hemisphere would be dominant (master) and the right would be subordinate to it. After all, one might think, look how useful the right hand is in comparison to the left. If you're right-handed, you probably use your right hand for anything requiring dexterity (hand writing, throwing a ball, painting, etc.). Surely the left side of the brain must be more developed and in turn the master. Needless to say, this theory assumes that language will always be in the same hemisphere that controls the dominant hand. However, as we will see, this is not always true.

The Birth of Cerebral Lateralization

After World War II, it was pointed out by several investigators that this rule of language dominance and hand dominance in the same hemisphere was not accurate. We soon learned that while right-handers would be almost exclusively language dominant for the left hemisphere, left-handers were also left hemisphere dominant for language for 60 percent of the cases. This would indicate an intrinsic and unique functional skill for each hemisphere. Therefore, hand dominance and language are not directly related. However, we can still learn about the brain by studying handedness of large populations in various skills. I will explain....

It has been well known that left-handers are seen more frequently in certain groups of gifted skills. These would include architects, engineers, and mathematicians. Could it be that by looking at the forest and not the trees, we can gain insight about brain function and this concept of functional asymmetry? By examining large groups of people and determining representation of handedness, we can gain insight about "cerebral lateralization" (the concept that left and right hemispheres are endowed with special skills). This really does appear to be the case. This form of investigation is sometimes called obtaining the *gestalt,* which is German for "form." It's similar to looking at the forest and not the trees. Sometimes insight can be gained by looking at the trees.

What's the Evidence for Functional Asymmetry?

With the help of modern technology, there has been an explosive growth of understanding of specialized areas within the brain and how they appear to be lateralized to the left or right hemisphere. This is because of the amazing new neurological imaging modalities that we currently have available, such as computerized tomography (CAT scan), magnetic resonance imaging (MRI), and positron emission tomography (PET scan). With these devices, we can demonstrate, with exact accuracy, where a lesion is within the living human brain and make the clinical/neuropsychological correlation that determines the functional necessity and lateralizing qualities of the area in question.

Modern technology has allowed us to understand our brains are further divided into areas of specialized skills. In reality, there is no master or slave hemisphere, but rather a lateralization or conservation between the two equal hemispheres, with each side having unique abilities. The human brain functions by the harmonious cooperation between the verbal (left) and the nonverbal (right) hemispheres. It now appears that the right hemisphere has specialized in visual/spatial skills, while the left seems to excel in temporal/verbal skills.

R. Duara and A. Kushch, in 1991, presented a good example of how these new imaging modalities can teach us about dyslexia. They measured six areas of the brain and compared them with the corresponding areas of the opposite hemisphere, using horizontal brain slices on MRI. They obtained brain scans from twenty-one dyslexic and twenty-nine normal control subjects. They discovered that dyslexic subjects seem to have a larger angular gyrus on the right when compared to the left. This area is located in the posterior temporal/parietal region and might explain the higher incidence of right-sided brain skills among dyslexic people.

This is not the only study demonstrating a difference in appearance in the dyslexic brain. In 1990, Hynd and Semrud-Clikeman compared the brains of ten dyslexic and ten normal controls, matched for age and sex, on MRI brain scans. They clearly found that a portion of the left hemisphere, planum temporal, was consistently smaller when compared with the right in the dyslexic brain. It is typical in the normal brain to have this portion larger on the left, and so this finding in the dyslexic brain is often referred to as reversed asymmetry. Thus, even in the developmentally abnormal brain there is anatomic evidence of a difference through equality.

Microscopic Evidence for Right-sided Visual/Spatial Skills

David Eidelberg and Albert Galaburda, from Harvard, reported a rather interesting finding in 1984. After an in-depth microscopic analysis of various areas of the cerebral cortex (architectonic parceling), comparing the left and right parietal lobes, a provocative discovery was made. First, there was marked asymmetry between the hemispheres (they appeared different). This was not very surprising. However, the area of the inferior parietal lobe called PEG had the most asymmetry. Interestingly, the right PEG looked very similar microscopically to the visually related cortices (occipital lobe). In other words, the right hemisphere looked more like the visual lobes than the left. This might be expected since the right hemisphere is believed to be specialized in visual/spatial skills.

Left Hemisphere Skills

The left hemisphere is dominant for manual skills, such as may be involved in eye/hand coordination, as well as verbal skills. A concert pianist reading music to play an instrument might be a good example of eye/hand coordination used by the left hemisphere. Regarding verbal/temporal skills, it would appear that the left hemisphere would be the hemisphere Sherlock Holmes would use to solve complicated crimes. Needless to say, one should have an excellent grasp of time (alibi) and language to "trip up" the suspect. If one were to believe that females have higher left hemisphere skills (as we will see in the next chapter), then perhaps females would be a natural at solving crime. Perhaps Mr. Sherlock Holmes should have been Mrs. Sherlock Holmes? Other skills of the left hemisphere include skills of communication. These are important in journalism, teaching, and the legal profession.

Right Hemisphere Skills

Let's take a case in point for the right hemisphere. While a neurological resident at Georgetown University, I saw a very interesting patient. He was a young man of about thirty years of age who had suffered a small stroke (infarction), damaging the inferior parietal area on the right hemisphere. He appeared normal otherwise. What is interesting is that he had been a very successful computer programmer. However, after the stroke, he found great difficulty in programming and generating a spatial concept of the solution. As you might be aware, computer programmers use a flow chart, a sort of graph, to display the algorithm of the various sequential steps involved in a program. No doubt the right hemisphere is used for this visual/spatial skill. Because of the lesion in the right hemisphere of his brain, this young man was no longer able to conceptualize a program in his brain. Thus, he had become disabled as a computer programmer.

Some believe that the right hemisphere is also the musical side. Others, such as Eidelberg and Albert Galaburda, believe the right hemisphere to be also dominant for what is called "arousal" and "vigilance" (to develop an interest and persevere until that problem is solved).

Have you ever noticed that during times of inquisitive emotional states, we often gaze upward and to the left with our eyes as if to search for a visual solution in the right hemisphere? Schwartz and Davidson wrote in 1975 that they felt this was more evidence for a right hemisphere dominance for affect or emotion. Many investigators consider the right hemisphere dominant for what might be called nonverbal emotion. This is still controversial.

Is Handedness Therefore Determined Prior to Birth?

Now one might say, I thought that handedness was determined after birth and that the probability would be fifty-fifty for either side being dominant, except for that society stigmata which favors right-handedness. It is true that in most cultures there is a history of bias that encourages individuals who would normally be left-handed to switch at an early age. One of my older Greek patients reported that as a child going to school in Greece he was forced to write with his right hand even though he was left-handed. He was told that the left was the devil's hand and was slapped on that hand with a ruler if he tried to write with it. All students were to write with the right hand. The result is that he is truly ambidextrous and able to write equally well with both hands. Even today in parts of central Africa children are often discouraged by their mothers from being dominant for the left hand.

What is interesting is the fact that this social pressure for favoring the right side is perhaps universal. This gives credence to the fact that the nature of things is normally for right-handedness to appear, that there is a biological basis for handedness and this basis makes most of us right-handed. This is not to say that a child could not be influenced to switch hand dominance (from right to left), since we all know that's possible. However, it appears that sometimes something neurodevelopmental takes place during fetal growth that makes a change from the more common right-handedness to left-handedness (around 90 percent right, 5–10 percent left). One could conclude that this congenital cause for left-handedness might be influenced by a more skilled right hemisphere.

Gesel and Ames help shed some light on this subject in their report in the *Journal of Genetic Psychology* in 1947. They discovered something interesting. They noticed that newborn infants, when placed on their back, would posture their head and trunk either to the left or right. Of the children who at the age of ten years were found to be right-handed, all (100 percent) had postured to the right in newborn infancy. It is important to note that of the 5 percent of the infants who postured to the left, all (100 percent) were to become left-handed. However, not all infants who postured to the right became right-handed. A small amount, 5 percent, who turned to the right became left-handers in childhood. Thus, if an infant postures to the left, there is almost a 100 percent chance he or she will be left-handed in adulthood. So there is some evidence for biological determination of handedness.

Where in the Hemisphere Is the Functional Asymmetry?

Where in the brain and which lobes are we talking of when we discuss cerebral lateralization? Is it the whole hemisphere that is different or is it just a part of the hemisphere? Results of studies done on monkey brains and later on human brain autopsies found that the cytoarchitecture (microscopic inspection) of the inferior parietal lobe seems to display the most anatomic asymmetry between the two hemispheres. In other words, only a small portion of the hemispheres was different. For that reason, when we talk of lateralization we can think in terms of asymmetries between the two inferior parietal and posterior temporal areas. This small area is often referred to as the planum temporal, and on the left hemisphere of most people it involves the interpretation of auditory language (Wernicke's area).

Conservation between the Hemispheres

What causes an individual to have a brain that is lateralized to specific skills (right or left)? This is the $64,000 question. There appear to be several conditions that could result in a brain with lateralized skills. It is known that surgical ablation (scarring) to an area of the fetal monkey brain causes the corresponding area of the cortex on the other side of the brain to develop a more extensive pattern of connections to both hemispheres (Goldman, 1978). In other words, there is some degree of plasticity to the hemispheres. If one side is wounded, the corresponding side attempts to compensate for the loss. It would appear that to some slight degree cerebral development between the two hemispheres is conserved.

Let me illustrate this. Dyslexia is a congenital defect in the brain that results in difficulty with reading skills. These people are not intellectually inferior or retarded. Their brain has difficulty interpreting written language. It is unfortunate that just a few decades ago these people were thought to be intellectually inferior and not candidates for higher education. I too was one and suffered from prejudice. Dyslexia was poorly understood. It now seems to be rather well established that these patients tend to have altered cytoarchitecture confined to the planum temporal of the left hemisphere. It is also fairly well established that dyslexics and their family members have a tendency to have superior right hemisphere skills. It is sad that with engineering and computer skills in so much demand, years ago these people were discouraged from attempting higher education. It would appear that in some of these individuals the brain compensated for the defect in left verbal/temporal skills by adding to the right visual/spatial skills. It's as if Mother Nature or the good Lord were trying to show compassion and fairness during development—different but equal.

I'm Left-Handed. Should I Therefore Be an Engineer?

When we discuss handedness and lateralization, we must first fully comprehend the statistical significance of our data so as not to come to erroneous conclusions. Just because, for example, we find among a mathematically gifted group a higher incidence of left-handers may in no way mean that of a group of left- and right-handers the left-handers would be expected to be superior in mathematics. It should be noted that some studies revealed lower scores in spatial skills among left-handed college students when compared to right-handers. Therefore, no one should be prejudiced on grounds of handedness. Let us not become biased or prejudiced. Never forget the phrase: *All squares are parallelograms, but not all parallelograms are squares.*

This means, logic does not always flow in both directions. Just because left-handers are more common in a given profession does not mean that if you are left-handed you should have the same unique skills. You should not assume superiority in right hemisphere skills simply because you are left-handed. By the same token, just because someone is right-handed, he should not deduce that he is great playwright material. This does not give cause for prejudice.

In 1924, S. T. Orton reported a very interesting discovery about children with dyslexia. He found that if the written material were read in a mirror or upside down, some of the children were able to read better. Obviously these children were not mentally retarded, so perhaps there was a problem with the left hemisphere's use of graphics. Norman Geschwind and Albert Galaburda developed an interesting theory of why these students improved their reading skills this way. What is the common denominator between reading upside down and in a mirror? It is that you would now be reading from right to left, that is opposite from the normal Western pattern. From our previous discussion about vision and the occipital lobes, we discovered that the left lobe is responsible for the right visual fields of both eyes. Therefore, as we are normally reading from left to right, the next letter or word will always be coming from the right visual field, which would require the use of the left hemisphere. Some dyslexics may require the superior skills of the right hemisphere for visual/spatial interpretation of graphics while reading. Therefore, when reading from the opposite direction (right to left), they are able to facilitate the better-developed right hemisphere for the incoming graphics.

When we discuss cytoarchitecture and cerebral lateralization, a question often comes to mind. Could we use microscopic examination to find out whether a brain is lateralized in skills to the right or to the left? There are some studies that measure the planum temporal and compare between left and right hemisphere. More work needs to be done. However, what is interesting is that the cytoarchitecture (microscopic examination) of the planum temporal is not only very different between hemispheres but also among individuals. Around this area there is a portion of the brain called Brodmann's area 39. Albert Einstein's brain was very interesting in this area. I will explain. . . .

Lay persons not familiar with neuroscience might assume that if one uses a portion of one's brain, it should generate more neurons and thus become more complex. If we exercise, don't we get more muscles? Well, it is true that when we use various areas of the brain for specific tasks, these areas receive more blood. We can see this on PET scans. It can also change in electrical activity, which we see on an EEG (computerized electroencephalography). However, we are only given a certain number of neurons in life, about 10 billion, and no more. What changes? There are two basic cells in the brain: neurons that transmit and store data and glial cells that act as metabolic support for the neurons. With constant metabolic activity, it is these glial cells that seem to generate to support the more active neurons. Einstein had a significantly higher glial/neuronal ratio or index in this area. He obviously

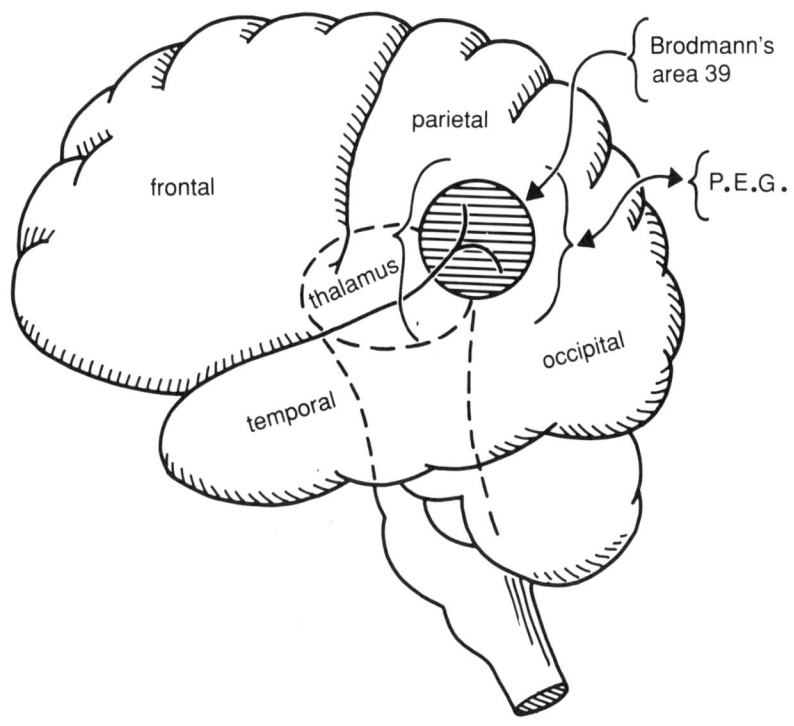

Fig. 8. Brodmann area 39.

needed more glial cells to support this more active and more complex area.

In Summary

The human brain is composed of a left and a right hemisphere. The hemispheres work harmoniously together, each contributing unique skills. The left hemisphere is superior in verbal/temporal skills, while the right specializes in visual/spatial abilities. We could refer to visual/spatial skills as nonverbal skills. There is no master/slave relationship between the hemispheres and therefore no true dominance. Both can work relatively independently of each other. Both are logical and creative in their own way.

Some persons may be lateralized to right hemisphere skills and some to left. Because there is no true dominance between the hemispheres, no one can say with any high degree of certainty that one individual lateralized to a particular hemisphere is superior or inferior to another, just different but equal.

The Skills

Left Hemisphere	**Right Hemisphere**
Temporal	Visual/spatial
Verbal	Music
Verbal logic	Emotion (initiation)
Arithmetic	Calculus
Verbal expression	Motor expression
Reading music	Playing by ear
Solving a crime	Fixing a clock
Reading a recipe	Following a map
Reading a novel	Building a model
Poetry	Sculpture
Painting a known landscape	Painting a landscape never seen
Legal case (presentation)	Programming a computer
Creating a novel	Creating music

2. Sex and Lateralization

This chapter looks at both scientific evidence as well as evidence from everyday life to explain differences in lateralization between the sexes. It provides insight into variations in thought patterns and behavior that reflect differences in brain lateralization (determined in utero) and describes sexual differences from a reference of functional equality.

Male and Female Differences

Before I entered Georgetown University to do my residency in neurology, I became friendly with a very fine older gentleman who was a psychiatrist at Prince George's General Hospital in Maryland. He knew of my previous career in the aerospace computer industry and my profound interest in the human brain. He told me of an observation he had made during his long career as a psychiatrist and thought that maybe some day I would make sense of it. He said that part of his mental status work-up was to test memory during the initial part of the evaluation of the patient. His habit was to give an address (number and street) and then require the patient to recall it after about five minutes into the interview. He discovered that females generally tended to forget the number, while males tended to forget the name of the street. I never forgot this, and as research began to unfold in brain lateralization this discovery by my most observant elder colleague began to make sense.

For many years, I have been known to collect directions from people coming to parties to see who has written them down in longhand and who has drawn a map. I have made an interesting observation that has seemed to be fairly consistent. If directions to a house, for a party or business, are given verbally, there is a fairly consistent way in which the female records the information as opposed to the male. For example, almost invariably the female writes it out in longhand (go to the traffic light, turn left, second house on the right, etc.), whereas the male fairly often draws a little map.

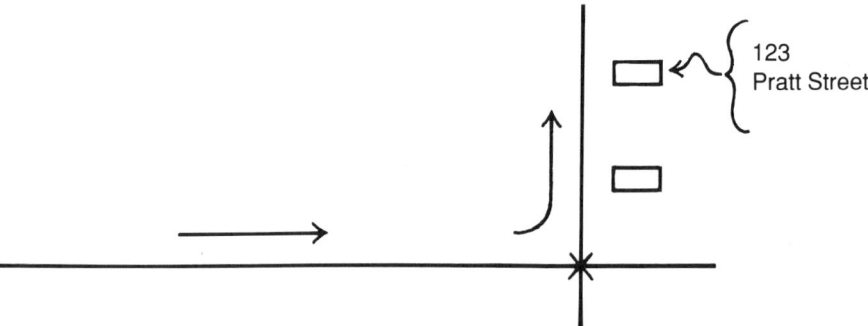

Fig. 9.   Drawing showing an example of visual lateralization typical of the male brain.

It would appear that the brain, when confronted with a task, lateralizes to the side where its skills are most adaptive to solving the problem. The same problem is solved, but the sexes each use a different solution. The result is the same no better and no worse; both find their way. This suggests that behavior follows skill, a skill that might have begun prior to birth.

I have an interesting story to explain this mechanism. When I was in training at Georgetown University, a girlfriend of mine telephoned for directions to an automobile carpeting company to have her car recarpeted. Of course, being one lateralized to the left hemisphere (female), she wrote all the directions in longhand. She asked if I could meet her there, but when given the paper with the verbal directions, I had great difficulty following them. She was perplexed for she thought it was very clear. She did not like maps. I told her to go in her car and I would follow in mine. She had no difficulty finding the place because of her left hemisphere verbal directions. My right hemisphere (visual/spatial) was able to create a mental map assisting our way home. There is no superiority or inferiority in this, just different ways of doing the same thing.

This is a classic example of two opposite hemispheres successfully at work at the same time and on the same task. When she was confronted with a task, little did she know that she mobilized her cerebral energies to the hemisphere for which she is most adaptive. In her case it was the left (verbal/temporal). She wrote her instructions in a way that would be simplest for her to understand, in the same manner in which she experiences her universe. Therefore, she wrote them in longhand. For her it worked just fine, for she had no difficulty locating the shop.

These are interesting clues to how different brains are able to solve the same problem. Life experiences are full of these clues and I could go on. However, are they truly meaningful or are they just stories without any significance or relevance? Is there any truly objective evidence of any difference in male and female brains?

Disorders and Sexual Lateralization

Going back to the concept of forest and trees, let's first examine the forest by looking at some interesting statistics. We have known for many years that several brain pathologies that appear to be secondary to disturbances of cerebral function in the left hemisphere occur in a predominance of males. These include dyslexia, autism, stuttering, etc. This would suggest, albeit not prove, that the nature of things regarding the development of the male brain may favor the right hemisphere more than the left. Some studies have gone so far as to suggest a significantly higher rate of dyslexia and stuttering among "strong" left-handers as opposed to "strong" right-handers (Geschwind). Although several studies indicate that left-handedness is more common in males, this does not seem to be confirmed at present.

Are There Structural Differences in the Brains of Males versus Females?

Now let's look at the trees. Is there any evidence of differences in laboratory animal brains between male and female? As early as 1971, Raisman and Field reported a sexual dimorphism (difference in structure) in rats in a central area of the brain called the preoptic area. This area is responsible for the release of certain sexual hormones. The male is larger and has more neurons in this area. Therefore, with only a microscope and this section of the rat brain on a slide, one is able to determine whether the brain is male or female. This is now a well-known fact and serves as a basis for many very interesting experiments.

Now one might say, "So they look different, does that mean they function differently?" Is there any evidence that a male rat and female rat might function differently electrophysiologically? The answer is yes. There is evidence of different brain functions between the rat brains sexually. It appears that male mice sleep differently from their female counterparts. This was just revealed by Fishbein and Fang in their doctoral program in neurocognition (1989). As you might already be aware of, during sleep we pass through several cycles of slow wave (deep sleep) and REM (rapid eye movements), which is the period in which we dream each night. It appears, based on this study, that male rats seem to have more REM sleep than females. It seems that within a species, there is a specific pattern of sleep that is measurably different between the sexes. One of the areas of the brain stem believed to control sleep is the pons. This discovery suggests that the pons is wired differently in different sexes. Thus, if even in the primitive portion (pons) of an old species (rats) there are electrographic differences between the sexes, how many more differences are there among humans?

Psychologists have known for many years that females will often respond physiologically differently than males to the stress of meeting a stranger. I have seen an actual demonstration of this most interesting phenomenon. Women will often have an increase in rate of respiration, with little change in blood pressure. Men, on the other hand, will have an increase in blood pressure, with little change in their respiratory rate. I am not sure of the significance of this observation, although there may be several plausible explanations. One possibility is the fact that it is the female who bears the stress of pregnancy. If one were an abdominal breather, typical of the male, the increased abdominal pressure caused by pregnancy would interfere with respiration, thus jeopardizing the safety of child and mother. Thus, the good Lord decided that women will be chest breathers and men abdomen breathers. Why is this important to our discussion? This finding suggests that the neurological center that controls respiration in the brain stem (medulla oblongata) is wired together differently in opposite sexes. Some believe that to be different is to be superior or inferior to another. However, cerebral lateralization teaches a difference through equality and utility.

Difference through Equality

It must be pointed out that, as mentioned in chapter 1, the concept of dominant and nondominant hemispheres is an old one, and as our knowledge of brain mechanisms increases, we find there is no master/slave relationship between the two hemispheres. The hemispheres work harmoniously together, combining the verbal and nonverbal worlds. The terms *dominant* and *nondominant* are still used in neurology but not quite in the same context as they were prior to the Second World War. Each hemisphere seems to be specialized in unique skills: the left, verbal/temporal, and the right, visual/spatial. Just because a person's skill may be more developed, be it the left or right hemisphere, one cannot assume overall superiority to another. One may be different but equal.

What about Freud?

This comes into conflict with Freudian theory. Freud felt that because males had a penis and females did not, this gave women a "genital deficiency"; they were thus not up to the equivalent of the male. He felt that the female mind was therefore unable to overcome "castration anxiety" and would forever be troubled by this. He believed that when a young daughter realized that her mother was also "ill-equipped" and responsible for bringing her into the world, she would therefore be driven emotionally to her father. Freud realized a difference between the sexes; however, neither sufficient technology nor scientific data was available at his time, which no doubt influenced his conclusions. Cerebral lateralization teaches that while there are differences between the sexes, they are not of an inferior or superior nature. We can be both proud and happy with these differences for they are degrees of lateralization (side to side) not a question of dominant versus nondominant. There is a tendency for the brain to be "conserved" in total number of skills between the hemispheres and, therefore, between the sexes. To be conserved means that the total number of skills is the same despite their distribution between the left or right hemispheres. If one were weak in left hemisphere skills, one can be encouraged to look for right hemisphere skills.

No one can dispute the importance of environmental influences on behavior (software). But how much is environmental and how much congenital (prior to birth)? Cerebral lateralization takes the congenital approach (hardware). There are differences that are hardwired into the brain. These differences are not wholly subordinate or dependent upon environmental or sociological influences. These differences may be a "fundamental law of nature" on which life depends for the survival of all mankind. Perhaps survival is the reason for the differences, an enhanced efficiency resulting from biological specialization. . . . To understand this is to understand the master engineer of our ecosystem. To know ourselves is to gain knowledge of God—for the human brain is the marvel of the universe and the crown of creation.

Dogs and Sexual Differences

My mother (now deceased) was extremely bright and observant. When my brothers and I wanted a dog, she would always suggest a female, saying a male would want to run all over town but the female would stay around and make a better pet. In our household we have had all kinds of dogs, male and female, and with every single pet she was always right. As a boy, and later as a man, I never forgot her observation. Later, I discovered that almost everyone I knew who was knowledgeable about dogs, including veterinarians, would agree with it. I began to believe that there may be subtle but powerfully significant differences between the genders in various species.

Veterinarians and other dog experts have told me that a female will make the best guard dog for the home. Although the male might be larger, the female is more "territorial." She will defend the home members with great energy.

Romance and Cerebral Lateralization

We have known for a long time the more obvious differences in the opposite genders. After all, you might say, little boys are attracted to little girls and little girls to little boys. There must be something hardwired into the brain that allows the opposite sex to attract. Pupil dilation studies were commonly done by psychologists back in the sixties and early seventies. These studies used a camera on the eye of the subject and measured the diameter of the pupil in response to visual stimuli. Some researchers have reported that they could tell if a subject was attracted to the opposite sex by observing their eyes for pupil dilation on gaze. However, there were some fairly consistent differences between the sexes in these studies. When we see something that stimulates us, there is a measurable dilation in the pupil of the eye. Men were attracted to the fairly obvious; bikini-clad girls, for example, caused pupil response. However, what was more interesting were the other differences. For example, men seemed to be more attracted to landscapes than were women, which might be explained by their right hemisphere (visual/spatial) orientation.

However, what is also very interesting is a scene that created great interest in the females. This was a sunset scene of a beach with a muscular man showing his bare shoulders standing alone. In his arms was a little baby. Could this be indicative of left hemisphere lateralization in the female brain? I say yes, and this is the reason. The left hemisphere is often called the logical side

Fig. 10. Picture of a man and baby on the beach to which women are drawn, indicating possible left brain lateralization in the female brain.

of the brain. However, that depends on the type of logic. Verbal logic, yes. Spatial logic? Perhaps not. I have always felt that a better description of this is that the left hemisphere is the practical hemisphere; this would agree with the concept of "verbal logic," which would indeed be practical. Neurologists have known for many years that stroke victims with only a left hemisphere were much more practical and functionally independent (eating, communicating, etc.) than those with only a right hemisphere. Is it possible that these practical instincts of the home and the immediate surroundings help the female dog stay closer to home, help the lioness protect her young, may also influence human female behavior in important practical aspects of life? For example, if you give $5,000 to an adolescent male, in twenty-four hours he will bring home a loud, gas-guzzling junk of a car, while a girl might think in more practical terms, such as saving for property.

There is a biological necessity for practicality in the female. After all, it is the female who gives birth and nurtures the young. One cannot ignore maternal instincts. This, I believe, is the common denominator. This is important for survival of the species. The two halves make a whole, and they contribute unique skills and values to that whole. These skills influence behavior, which, in turn, assists in our survival. This may be more obvious than you think. Next time you and your spouse grab the newspaper, watch. Often the male will read world news, while the female will be attracted to more practical news in the local area. This has often been my experience. Perhaps you will notice this as well.

Of course, we must remember this is not necessarily absolute (differences exist within a sex). However, the difference is, perhaps, everything. The difference is what makes a marriage. The two halves make a whole and by virtue of the unique contribution that each member makes, the whole (male/female unit) is greater than the parts and thus, being acceptable to nature, is blessed with offspring. There is no superior or inferior, as earlier theorists proposed, just differences that may assist with the survival of the species—biosexual specialization.

I look upon marriage as a wheel to a car. The male represents the outer tire and the female the inner hub. Together a more functional unit is formed. The wheel becomes "whole," and the world or romance goes on.

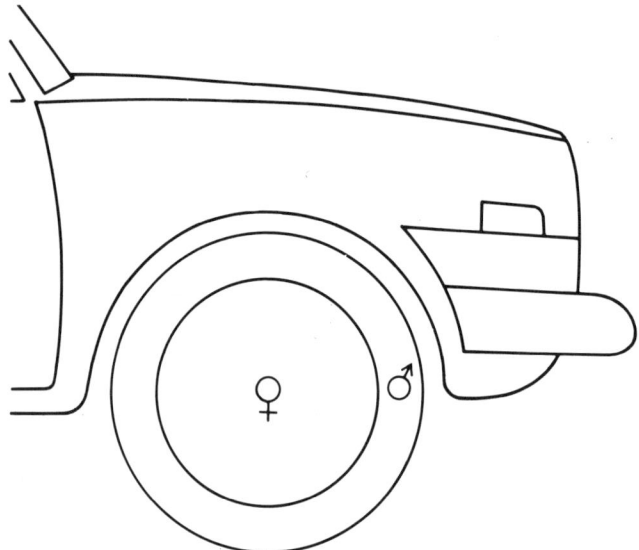

Fig. 11. Marriage as the wheel of a car; the male represents the outer tire, the female the inner hub.

Sex, Lateralization, and the Developing Brain

Scientific Data

Now some may say that all this sounds poetic, but are there any more scientific studies that suggest a relationship between sex hormones and cerebral lateralization? Nass and Baker, in 1987, reported on an interesting study that indicates a relationship between sex hormones and handedness. I will explain.

There is an interesting medical condition called congenital adrenal hyperplasia. This is a defect in the adrenal gland, which sits just above the kidneys. The defect is a lack of enzyme 21-hydroxylase, which causes a feedback inhibition of the hypothalamic-pituitary-adrenal axis. The result of this inhibition is an excess production of the male sex hormone testosterone. This disease occurs in both boys and girls. The theory was that if sex influences cerebral lateralization, then perhaps sex hormone secretion in utero (unborn child) could be a causative agent in handedness. If so, then as there appear to be more left-handed males than females, which indicates a right hemispheric lateralization in the male, would there not be an abnormally high incidence of left-handedness in females with this disease; after all, these girls would be exposed to abnormally high concentrations of testosterone.

A survey by mail, using what is called the Edinburgh Handedness Inventory, was sent to 120 families in which at least one member had this disease. Using this data, researchers developed a Laterality Index. The result of this study was that female patients with this disease showed a significantly greater left-based Laterality Index than their normal sisters. Thus, it would appear that male sex hormone secretion in utero has an important influence in handedness and therefore, perhaps, cerebral lateralization as well. It also suggests that handedness is biologically determined prior to birth, long before it is expressed later in life.

The Tomboy

This disorder also revealed another interesting discovery. Young girls with congenital adrenal hypertrophy display different behavior patterns from their normal girlfriends. Money and Ehrhardt, in 1972, reported an increased tendency for females with this disease to play in rough sports, similar to little boys. Apparently the girls were normal in every other respect. Could this be a key to understanding the "tomboy"? Here again, there is evidence that much is determined in utero. Could it simply be the timing of the phase and/or amount of fetal hormone secretion (in this case testosterone) having a powerful effect on behavior and personality later on? Thus, even within a gender there are various differences that might be predetermined in utero. Some girls are more like boys than others. Could this not be another expression of difference through equality? Thus, it is possible that the tomboy you knew as a child might have been predestined to be a tomboy prior to birth. This causes one to wonder how much more adult and adolescent behavior has its root in the uterus?

Hormones and Receptor Sites

The evidence is now clear in modern neurobiology that there are differences between the male and female. We all know about hormones. These are chemicals produced by the body that travel to other parts of the body and have a particular affect on the part that is unique to the type of hormone. For example, the male sex hormone testosterone is produced by the testes. This hormone is responsible for the physical male characteristics. They include male-pattern baldness and hair growth in areas not typical of the female, such as arms and chest, etc. The female hormone estrogen is responsible for the female sexual characteristics, such as breasts, etc. The function of these hormones is to mold the male and female bodies and give them the sexual and physical characteristics of a male or female (vagina, testes, breast, muscles, hair, etc.).

Located throughout the body are various receptor sites at particular locations on the body. These allow the hormones to manifest their characteristics at a particular site. For example, on the breast there are ample receptor sites, which are small molecules on the cell wall, where the female sex hormone can bind to allow that cell to contribute, in its small way, to the formation of the breast. We have known of these sites for hormones for many years. However, what is relatively new is the fact that there are also sex receptor sites in the brain cortex of the newborn rat.

Why is it interesting that sex hormone receptor sites can be found in rats' brains? Two important facts must be kept in mind: (1) they are found in newborn rats, which are gestationally equivalent to a female fetus in late pregnancy, and (2) they are found in the cortex of the rat brain, and, as you may recall from chapter 1, the cortex of the brain is responsible for consciousness and thought. Could it be possible that these receptor sites function to allow hormones to affect sexual characteristics in the brain at the time of development, allowing a male brain or a female brain to express themselves? Thus, there would be characteristics unique to the sex of the animal in the cortex that might affect behavior and thought patterns. This is rather profound. Let me give you an example.

I remember growing up as a child and being fascinated about the differences between boys and girls. There was a rather interesting event in my life I will never forget. We were a group of little boys and girls with an adult guardian in a hospital. I recall our female guardian telling us that if we were to behave, we would have the pleasure of seeing the babies. Well, I observed that while the boys were less than enthusiastic, the girls were all excited about the idea. The boys thought this was unfair for they would prefer something else. Our guardian could not understand the dilemma. However, from that time on, I was aware of the difference. The boys were simply not interested in seeing the babies. No way am I convinced that this was sociologically induced at the age of five or six. The brains of boys and girls were just hardwired to be attracted to different things.

What is the significance of this story? It is now known that sex hormone receptors, estrogen in particular, are also found in the cortex of the infant monkey. What is striking is that they are more frequently found in the association areas of the brain. If you remember from the first chapter, these areas of the cortex are responsible for interpreting information that enters the brain. This information could be tactile (touch), visual, or verbal. Perhaps males and females see and hear a different world, starting very early on in development. We look at the same object and get a different impression. We may see the same thing but from a different point of view. This may be why we have a male cologne and a female perfume. Picture this, a man whispering "sweet nothings" in a woman's ear, and a beautiful woman gently pulling up her dress six inches above the knee. One picture, different reactions unique to the sexes. The young girl starts feeling warm below while the boy starts wondering if his pants aren't a bit too tight.

Sex Hormones and Sexual Desires

Now some may argue that this proves very little. Special microscopic receptor sites in the brain cells for sex hormones do not automatically mean that the brains are of themselves different. You might say, "I thought the sexes were different because of different sex hormones flowing in the blood of the body, testosterone in the male and progesterone/estrogen in the female; that these hormones influence the human brain in such a way that males are attracted to females and females to males? The brains are actually the same, but because they are under different hormonal influences, they result in different behavior." I will explain why this concept is probably not true.

We have known for many years that if you introduce male hormones (androgens) to females, they do not change their heterosexual preference. They will grow body hair similar to a man's. They may slowly develop a malelike physique. However, they do not change their sexual preference. What often happens is that the same sexual desires are enhanced or aroused. The female becomes more aggressive in the pursuit of sexual relations or intercourse. It is not the direction that changes; the sexual orientation or preference is the same, only the degree of aggressiveness changes.

A similar thing happens in the male. If a male has an increase of estrogen (which can occur in liver disease because the normally small amount of estrogen in the male body is broken down by the liver), he will not change his sexual preference, but there will be a reduction in sexual desire. Again, same direction, different degree of aggression. This concept is well known to doctors involved with alcoholic patients with significant liver disease (decreased libido). So one may conclude that to enjoy sex, drink less. However, the significance is that it is not one brain responding differently to different sex hormones but rather two brains of different sexes responding in different directions to the same hormone. Therefore, something as old and well known as this phenomenon lends credence to the concept of a different neuronal structure, or hardwiring, between the sexes.

A good way of explaining sex hormones and sexual preference

is a documented story I have of a geologist who became a male-to-female transsexual. Although he had experienced a successful marriage with children, he had this gradually overpowering desire to become a woman. He described it as a "woman inside dying to come out." To the shock and dismay of family and friends, he made his announcement, which terminated his marriage. He then underwent surgery and hormonal therapy and started to become female.

However, what I found interesting is the fact that he denied having any homosexual tendencies prior to his sex change. He clearly stated that he was not homosexual, but rather, a gender dysphoric. Furthermore, and more enlightening, is the fact that during female hormone therapy (estrogen), he developed an interest in males. He started to notice "cute guys." Clearly, this is not a typical response of a male brain to female hormones. This is, however, a typical response of a female brain to a female hormonal environment. Thus, he had female brain receptor sites and, therefore, truly was a female brain in a male body. On observing this, I was pleased that he had discovered his true identity, and I wish *her* all happiness.

Female Rat Brains in Male Rat Bodies

So, one may say, this is very interesting; however, are there any objective neurobiological studies that indicate a difference between the sexes? Geschwind and Galaburda described a very interesting neurobiological phenomenon in the rat brain (*Journal of Archive of Neurology,* 1985). It is known that the female rat brain differs microscopically from the male in an area called the preoptic nucleus. Also, the rat brain at birth is equivalent to the human brain in late utero. The human male testes is developed in utero and during that time there is a burst of testosterone. There is yet another burst of this same hormone much later during puberty. In the rat, however, this first burst of testosterone occurs during the development of the testes just after the rat is actually born (within fifteen days). Thus, if the male rat is castrated (testes resected) within the first two weeks of birth, then these potentially male rat brains would not be exposed to that first burst of testosterone. It was soon discovered that the brain of these castrated male rats would change into a female brain microscopically, despite receiving testosterone injections after birth. It is interesting to note that even the cyclic hormonal responses (gonadotropin) would be female. Thus, a rat brain that would normally be male could be changed in the laboratory to the other sex morphologically, at least in regard to the region of the brain called the preoptic nucleus. Could this be a key to understanding the development of the male homosexual?

How about the reverse? Can we go from female rat brain to male rat brain? I'm sure you know the answer is yes. We also know that if the female rat is injected with testosterone within two weeks of birth, the same area of the brain, the preoptic nucleus, becomes male even though the rat is female morphologically. Could this be the key to understanding the biological mechanism in the development of the lesbian brain?

It appears that testosterone has different effects on the brain at different times in development. There are two doses of this hormone. The first dose in utero and its timing might affect the sex of the brain. The second dose in puberty might influence the sexual drive. First dose for direction, second dose for aggressiveness.

This reminds us of the famous biblical text taken from the Book of Samuel: "Man looketh on the outward appearance but the Lord looketh on the heart" (Sam. 16:7).

Thus we see that there is evidence of a biochemical basis to the development of sexual differentiation in the human brain. That during development, the brain is exposed to various chemicals, and these chemicals (hormones) appear to guide the developing brain to either a female or male brain. It is as if this first burst of hormones is to determine further cerebral development, whether this is to be a female or male. Then the second determines intensity. It is a well-known fact in neurology that the brain is not fully developed at least until the twelfth year. At this age, the EEG (electroencephalogram) shows a normal adult posterior dominant rhythm of 8 hertz (cycles per second) or more (alpha range). Before this age, the normal brain will be slower, and gradually increasing in frequency from delta (2–4cps) through theta 5–7cps) and then alpha (8–13cps) at twelve years. Could it be that nature says that since the direction (sex) of the brain has been given in utero, let's give it some intensity now that it is developed and ready to accept sex hormones in puberty. Thus, girls and boys start looking at each other a little differently at that time.

Differences prior to Puberty

Now, some critics may say, "OK, I accept that the brains might be different in different sexes. However, the fact that boys and girls are attracted to each other during and after puberty is not exactly new. What about before puberty? If the sexual brain is determined in utero, are there indications of differences prior to the age of twelve?"

I have made an interesting observation over the years. The next time you see a children's program, study the commercials. Business is all too often quick to learn where academia is slow. That's why I have always been a firm believer in the free enterprise economic system. Never underestimate the power of individual incentive. If there is a possibility of profit, business will find a solution to almost any problem. More commonly, business will find a more efficient way of doing things. For example, businesses learned long ago that to increase the volume of sales they must maximize the volume of media exposure in their advertisements. In other words, advertise to all potential customers as much as possible. You will notice that if a toy company has been allotted thirty seconds of advertising space, it will delegate fifteen seconds to a female toy and fifteen seconds to a male toy; fifteen seconds goes to Barbie Doll and fifteen seconds to GI Joe. They have learned that during this primary stage of sexual development (before puberty), there is a marked difference between the sexes in audio-visual appeal, and that the masculinizing brain will respond to one stimulus while the femininizing brain responds to another. Thus, the entire field of potential customers is covered and profits are maximized.

Taylor in 1969 reported, in a study based on children with febrile seizures who later developed temporal lobe epilepsy, that there is a difference in growth rate between the hemispheres in the sexes. He concluded: (1) that the left hemisphere matures after the right, and (2) that the male brain matures after the female. This would explain why boys often have difficulty with the typical left hemisphere subjects of reading, writing, and arithmetic in early education, which suggests that boys might be better served if they started school a year later than girls.

Morphologic Data for Sexual Cerebral Lateralization

Now a question may come to mind. Is there any morphologic (microscopic) evidence that would help us understand sexual lateralization? Is there any way we could look at the brain and see objective evidence that the female brain might have skills lateralized to one side and the male to the other?

In 1981, Diamond, Dowling, and Johnson reported an interesting observation in the rat brain. They measured the thickness of the cortex of the rat brain in specific areas of the posterior regions and compared these corresponding regions between the left and right sides. (As you may remember in chapter 1, the cortex is the crust or outer shell of the brain and is responsible for complex thought and intellectual function. This area is often referred to as the cortical mantle.) They found in the rat brain at birth that specific areas of the female were thicker on the left when compared to the right. Of course, the male brain showed specific areas of increased thickness on the right when compared to the left at birth as well. What is interesting is that later in life this lateralized asymmetry revealed itself in the mature rat brain as well. This finding suggests that the female might have special skills associated with the left hemisphere and the males with the right. I call this a gender specific cerebral advantage.

I would like to point out an interesting observation of mine during my previous career with aerospace computers. The men, although extremely talented in calculus and physics, would often be lopsided in their abilities, showing fewer left hemisphere talents. The women engineers, on the other hand, tended to be overall balanced, bright, and observant. One could say that they were global in their cerebral skills. They were not only good engineers (right hemisphere), but their communication and spelling skills were excellent (left hemisphere). This suggests that if the right hemisphere is so good in the female, then so is the left, thus resulting in an engineer with excellent spelling skills. I call this global-hyperfunctioning "superbrain"). In contrast to this, there were many brilliant male engineers, who were lucky just to be able to spell their names correctly. Needless to say, I have to include myself with this lot, for I also have poor left hemisphere skills in relation to the right.

Sensory and Motor Differences as Well?

There is even some evidence in the rat brain between anterior and posterior (front to back). Some areas in the front portions of the brain in the male appear to be consistently thicker, as opposed to thicker areas in the posterior of the female. This still needs to be examined, but if it turns out to be true, it will suggest that the sexes live in a different motor and sensory world as well. It would present some interesting thoughts on how the different sexes might view their worlds. The front is known to be the motor half, while the posterior is the sensory portion of the brain. This would suggest that the female would be more sensory oriented while the male is more motor oriented. Could this be the reason why women throughout the world are fond of mink coats and silk garments? I have often found it interesting to watch ladies in clothing stores finger items between their index finger and thumbs while their husbands stand with their hands in their pockets. Perhaps you have seen the same. It would appear that women might be more sensitive and are able to acquire more "information" in this manner than the male. Now males may finger the clothing also, but the ladies seem to finger them longer and also may ask a lady friend to do the same. Could this also be why the male is often more adept at throwing a ball? Perhaps he is using a combination of his specialized skills in the frontal (motor) and the right hemisphere (visual/spatial) to place the hardball with speed and accuracy.

Why the Ocular Dominance Test?

Ocular dominance is of some importance to cerebral lateralists. It has been said that the eyes are the window to the soul. I am not sure of the significance of this or its relationship to whether one is left or right hemisphere lateralized. There have been no conclusive studies on ocular dominance and lateralized skills. This should be done, and I hope that by including the chart with this book, studies will be started. There have been some studies regarding its relationship to handedness, which I don't feel have been relevant or conclusive. However, I will explain the concept and perhaps you might decide for yourselves.

The idea is that even though human vision is binocular (see objects with both eyes), many if not most of us use one eye as our primary visual receptor. The other we use simply for depth perception. The theory is that because some of us have brains that are primarily lateralized to either the left or right, this higher-developed hemisphere draws more visual data for interpretation, which in turn causes the contralateral (opposite) eye to become dominant. Even though we see with both eyes at the same time, one eye is dominant for primary vision and the other for depth perception. Example: If one were left eye dominant, could one therefore be said to be skilled in the right hemisphere? Another example would be in a female. Would left eye dominance (indicating right hemisphere skills as well) imply global cerebral functioning? Nobody knows. Let's see what you are. You can find out by using the chart that comes with this book or simply follow the directions below.

How to Perform the Ocular Dominance Test

I presume there is a door in the room that you are now sitting

Fig. 12. Women demonstrate their sensory orientation by feeling fabrics.

in. Open the door just a little so that the edge is showing. Now, with a pen or pencil held in front of you in your outstretched hand so that the pencil is vertical (up and down), using both eyes, try to line the pencil up with the edge of the door from across the room as best you can. Now close your *right* eye. If the pencil is still lined up with the door, you are probably left eye dominant. If the pencil swings to your right of the door edge, you are probably right eye dominant.

You can also try the reverse. Line up again with both eyes. Now close your *left* eye. If the pencil swings to your left of the door edge you are most probably left eye dominant. If the pencil stays lined up with the door edge then you might be right eye dominant.

You might then want to think again and see if your eye dominance matches what you feel your lateralization of skills might be. Needless to say, one would expect a left eye dominant person to be right hemisphere lateralized in skills and a right eye dominant person to be left hemispheric. See if you match. I am right-handed, but perhaps because of my profound right hemisphere lateralization in skills, I am strongly left eye dominant, even though both eyes have exactly the same refractive error. However, it must be kept in mind that this test is in no way absolute and there might be many opthalmalogic variations that could interfere with its reliability for forecasting cerebral lateralization. But you might find it interesting at social gatherings with friends. In chapter 8 I describe the left and right brain person.

Sexual Lateralization in Proper Perspective

Thus the hemispheres are balanced overall in the human race. The female tends to contribute skills from one side and the male from the other. There is no question of dominance or inferior versus superior but rather lateralization (side to side). Cerebral energies appear to be conserved between the sexes. However, there may be a gender specific advantage. For example, a female with good right hemisphere skills or a male with good left hemisphere skills might be more global in cerebral functioning (a

Fig. 13. The Ocular Dominance Test.

superbrain). This must not be misconstrued. Just because one is of a particular handedness or sex does not automatically mean superiority in a specific skill over another. Remember, *All squares are parallelograms, but not all parallelograms are squares.*

The logic does not flow in both directions. Even though there may be tendencies in a given sex population, one could not conclude that Jane will have to be a better speller than John or John a better engineer than Jane. There could be vertical as well as horizontal differences. Therefore, there should be no prejudice between the sexes. If Jane wants to be an engineer, she should be encouraged, for she could be globally hyperfunctioning (a superbrain). Instead, we should use this new knowledge of cerebral lateralization to help us understand our friends of the opposite sex.

Difference and Understanding

So there are actual cerebral differences between the sexes, not superior versus inferior differences but rather left versus right—different but equal. Both are creative and logical in their own way. These differences allow the sexes to look at the same thing in different ways and get another emotional response. These differences are often determined prior to birth. They may be hardwired in and not easily changeable. These differences should not allow us to be prejudiced on grounds of sex; Jane and John should be whatever profession they want to.

One may have different needs and desires. Often one looks at one's mate and wishes he or she were more like oneself. Perhaps we forget that it is the difference that attracted us in the first place and that this difference may be hardwired in long before birth and our spouse is not likely to change. Lateralization should teach us tolerance. A mistake often made in relationships is when mates constantly try to mold the other in their own image. They try to minimize the difference. Why isn't he more emotional? Why isn't she more assertive or interested in sports? Perhaps we should just enjoy the differences and let men be men and women be women. Often what happens during this molding process is that one or both of the sexes find they are losing their sexual identity. She hates to see him watch football or those war movies. He hates to see her watch those sitcoms. A very important key to a successful marriage or relationship is the ability to allow the opposite sex the freedom to express their maleness or femaleness. Perhaps we should encourage our wives and husbands to spend time with friends of the same sex, husbands with male friends, wives with female friends. Enjoy the differences and don't fight them for they are not ones of superiority but of equality.

3. Homosexuality

As a child, I was very fond of "The Three Stooges." I must admit that even today I find their skits rather funny. I remember the beginning of one movie in which there were two professors facing each other. One kept saying "environment" and the other "heredity." Needless to say, the comedy was about whether being a stooge was caused by environment or heredity. The thought was that if idiocy was caused by environment, it could be changed by environment to normal behavior. A bet was made, and, of course, in the case of the Three Stooges, heredity won.

Now that's a great story, but what do the Three Stooges have to do with homosexuality? Surely, homosexuals are not idiots. Homosexuals are just as bright as anyone else. But this little story deals with the controversy of environment versus heredity, which may be as old as history. This is not to suggest heredity in a strict sense for homosexuality as if it were recessive or dominant in transmission. Parents with a homosexual son do not have a higher risk of homosexuality with their next offspring. The word *congenital* might be better. For this describes possible neurobiochemical/environmental mechanisms that may dictate "sexual preference" beginning in utero (prior to birth).

The argument for congenital versus environment for homosexuality is also very old. I remember as a neurology resident at Georgetown arguing the many theories on this very subject with a fellow resident. However, before we get into the discussing research and biochemistry, I would like to review some of these theories with you.

Theories about Homosexuality

Basically, there are four views or theories on the origin of homosexuality. They are as follows:

1. Psychodynamic Theory (Freud). This theory states that the cause of male homosexuality is secondary to an unresolved Oedipal conflict. The male for some reason was unable to grow out of his attachment to his mother and then, later, to identify with his father. Thus he was unable to achieve a mature stage of "genital sexuality" and, in turn, suffers from "acute castration anxiety." He therefore avoids female sexual contact for fear of injury to or loss of his penis and for that reason becomes a male homosexual.

2. Behaviorist Theory. This theory describes homosexuality as a "learned sexual preference." The cause could be from a demanding mother or father or other oppressive female. Thus the young male would be pushed into homosexuality and mothers not desiring homosexual sons might blame themselves. Others might call this the "try it, you might like it" theory of homosexuality.

3. Humanistic-Existential Theory. This theory holds that the homosexual had no "authentic choice" in the matter of sexual orientation; that it is a pattern of "disturbed functioning" which is causative. In other words, "that's the way it is, so deal with it" or "don't fix it if it is not fixable" theory.

4. Biological Theory. This theory states that gays are not gays by choice but by an inborn developmental change. Some felt that it was due to chronic sex hormonal deviation from the norm, but this was later proved not so. Others felt the change occurred in utero. One might think of this theory as a "female brain in a male body" causative factor although this might be an oversimplification. The cause may be more organically (hardwired) complex than that. This theory has lost favor in the last several decades possibly because of the advent of behaviorism. However, as we shall see, its disfavor might have been a bit unfair.

I grew up believing, as many others, that homosexuality was in some way caused by environment. There were many theories and "old wives' tales." Some believed that men became homosexuals because of dominant mothers. Others felt it was secondary to abusive mothers or older sisters. Another theory was sexual experimentation; that because of curiosity and experimenting with sexual adventures with the same sex, one could become homosexual. There may be some truth to all of these theories but as a dominant cause I seriously have doubts. One of these doubts is the term *sexual preference*. That is to say, at some stage in our development we might have said: "Well, last night I had a wonderful time making love to a woman. I think tonight I will try making love to a man." I doubt that most of us achieve "sexual preference" through experimentation or sampling.

There is one wrench in the gears of this environment theory that was of great concern to me. It is this. Gays are gay in spite of the great and unjust barriers that are often thrust upon them. Especially in the fifties and sixties, these barriers were rather strong. That gays are gay in spite of tremendous social pressure to be the opposite makes it appear that they are gay because of some quirk in the social system rather than to spite the system. It is as if there is some very powerful inner force that allows them to express themselves in the sexual manner they truly desire and to overcome these colossal social barriers. We are learning every day that much more is determined prior to birth than we ever dreamed of. As more knowledge is gained, there is a congenital trend in science today of which schizophrenia is a good example. I will explain.

The Congenital Trend of Scientific Thought

Schizophrenia is perhaps the most common psychotic disorder. Almost 1 percent of the world's population suffers from this problem. This condition was first described by the German psychiatrist Emil Kraepelin (1856–1926), who felt that the defect that caused this problem was hardwired into the brain and not socially induced. At that time he called it dementia praecox. Later a Swiss psychiatrist, Eugen Bleuler (1857–1939) changed the name to the current schizophrenia. All this time it was felt that the problem was not caused by the environment.

However, years later Sigmund Freud (1856–1939) started

Fig. 14. Can one become homosexual out of curiosity and sexual experimentation with the same sex?

the psychoanalytic theory and most of psychiatry of that day thought this disease was caused by environment and was a manifestation of a "weak ego." Freud did not have the scientific advantage that we have today. However, today we know that schizophrenia is clearly the result of congenital (genetic in this case) causes and not the result of environmental influences. We learned this by discovering that the prevalence of this disease is between 10 and 15 percent among first-degree relatives. Also, many studies show without a doubt that there is a significantly higher concordance rate (occurrence of a given trait in both members of a twin pair) for schizophrenia among monozygotic (identical) twins than among dizygotic (nonidentical) twins of the same sex, even if reared in different families. Freud did not have this information and therefore concluded incorrectly.

I have selected schizophrenia as an example not to suggest homosexuality is a disease (for it is clearly not) but because it also may be involved in cerebral lateralization. In 1977, C. Boklage reported that the risk of schizophrenia may be associated with the process of cerebral lateralization. Needless to say, this is very interesting. He found that the concordance in monozygotic twins was close to 100 percent if the twins were right-handed. However, if one or both of the twins were not clearly right-handed, this concordance fell markedly. Thus we now see that schizophrenia may not only have its origins in utero but may be caused by the same process that causes cerebral lateralization. Perhaps a defect in the right hemisphere of the brain. We now realize that as neurological research unfolds, adult behavior often has its roots in the uterus.

Now let's look at homosexuality from a casual observational point of reference. Most of us have known homosexuals, and I would like to give some anecdotal observations, in some cases as their friend, in others as their physician.

Personal Observations of Homosexual Behavior

I recall many years ago as a medical student about to do my first hospital rotation in the Washington, D.C., area, finding affordable housing that would be safe and near the hospital was difficult. For that reason I was often bugging the Education Department for any news that might indicate available housing nearby. One day I seem to have become lucky. The director said that there was one place that was near the hospital, safe and very affordable. It was a nice, three-bedroom apartment with two available rooms. This was perfect. "However," the director said, "there's just one catch. The student in the third bedroom is gay and outspoken; therefore, filling these rooms has been fairly difficult." This might be even more surprising, since this was before all the AIDS hysteria. Being more curious than fearful and not being gay, I thought this might indeed be most interesting and so, of course, moved right in. He was in several ways a real credit to the word *gay*, always being very pleasant and a true lover of life. I remember the first time we met. He said, "Hi, my name is M and I'm gay." I simply replied, "My name is Dave and I'm not." We both respected each other and got along very well.

We soon made an agreement that M would do nothing to make me feel embarrassed. Actually, we became rather good friends. I soon discovered that he was very open about his sexual preference as were many, if not most, of his friends. They would openly and liberally discuss with me all the interesting facets of their sexual histories. These interviews allowed me to come to some fascinating conclusions:

1. Almost invariably, gay men first started to realize that something was different around the age of puberty. As discussed in the previous chapter, this is the time when there is another burst of sexual hormones that would give "speed" to the already predetermined sexual direction.

2. Gay men often seemed more promiscuous. This was made very evident to me while a medical resident at an out-patient clinic just when AIDS and its relationship to promiscuity was becoming well understood. Some of my gay patients would say, "Doc, I know about the odds; I'm down to an average of fifty sex different partners a year." Not all displayed this extreme behavior, but certainly many were more active than most heterosexuals I have known.

3. Not all gay men prefer men with effeminate manner-

isms. There was another gay medical student, P, who was very effeminate, sometimes to the point of wearing makeup. My friend M made it clear that he was not attracted to that kind of man. He wanted his men to be men. He stated that he never had sex with P simply because P was not his type.

My friend would go out almost every night and play volley ball. Afterwards, he would pair up and spend the night at a different place with a different lover. He made no secret of his life-style and would defend it strongly if pushed.

This is not new. Saghir and Robins (1969) reported that male homosexuals found it easier to separate sex from emotional involvement and thus could engage in a higher incidence of promiscuous activity. Female homosexuals (lesbians), on the other hand, tended to be much more monogamous in their relationships.

This is not to say that all male homosexuals have sexual appetites that rival rabbits. But anecdotally, they certainly do appear to be more active than their heterosexual counterparts.

A Biological Explanation for Homosexuality

Could there be a biochemical explanation for this behavior or is it simply socially induced and, if exposed, anyone could be affected? Do parents who want their children to be heterosexual need to worry about the influence of gay teachers on their children? Or is there a scientific explanation, suggesting homosexuality might be determined in utero?

Female Rat Brains in Male Rat Bodies

As you may recall from the second chapter, there is an area of sexual dimorphism in the rat brain. This area, the preoptic nucleus, is microscopically different between the male and female rat brains. In the male, it is larger and contains more neurons. It is also known that the rat at birth is gestationally equivalent to the human fetus in late utero (several months prior to birth).

During this particular stage, the male rat testes are just starting to become developed and begin secreting testosterone shortly after the first two weeks of life.

In the human male the testes become developed and start their first burst of testosterone while still in utero. Now, as reported in our second chapter, if the testes of the newborn male rat were cut (castrated) so as to prevent that first burst of testosterone, the preoptic nucleus of the brain would be morphologically female later in adult life, even if puberty was duplicated by testosterone injection later in life. The interesting thing is that even though the rat would be phenotypically (body-wise) male, this portion of the brain, if not the whole, would be female. It is also interesting to note that this area (the preoptic nucleus) would cyclically release gonadotropin (hormones) as if the body were female. Thus it would appear that we would have a sexually female brain in a male rat body and that gender orientation of the brain may be dependent on the sex-hormone environment during the early stages of brain development. In other words, the first shot of hormone prior to birth determines sexual orientation; the second, during puberty, determines sexual velocity.

The literature is now full of experiments regarding behavioral feminization where animals phenotypically male would mount males and respond sexually to male pheromone (odors). Could these animals be considered homosexual? Rats are a long way "down the trail," phylogenetically, from humans. However, it is the predictability of these experiments that make them remarkable and suggest an intrauterine biochemical mechanism for the human male homosexual.

However, it is interesting to note that some of our German colleagues have reported an increased incidence of male homosexuality in those Germans born during or shortly after World War II. The hypothesis is that stress in pregnancy might account for the environmental biochemical changes that might favor homosexuality. For that reason animal studies have been performed where pregnant animals were confined to small cages so as to induce stress. The result of these experiments seem to show an increase in frequency of "homosexual" behavior in regards to mounting and response to sexual odors (G. Dorner, 1983). Could stress or other factors in pregnancy biochemically shift the sequence of events that take place during brain development and thus render a change (hardwiring) of the brain sexually? Could it be that there is a "window" during human fetal brain development that would dictate that if testosterone or another hormone or biochemical is not available, the brain may have a "feminizing" orientation?

If this were true, then gays would not be guilty of "sin" and might not be any more guilty of homosexuality than straights are of heterosexuality. They would have no more control over their homosexual drives than the heterosexual does over his or her drive. They would simply be acting out their intrauterine biochemical sexual predestination. One might even say that, from a pure neurological point of view, theirs is not a homosexual act but rather heterosexual, for it is still a female brain attracted to a male brain. This is not to negate social influences of environment on behavior. But as I pointed out with the history of schizophrenia, to ignore biochemical influences on human brain development might of itself be prejudicial. As knowledge of brain function grows, biochemical causes of behavior become evident.

Ward and Weiz (1980) made a profound and interesting discovery. They found that if a female rat was stressed during pregnancy, there would be a rapid rise of the male sex hormone testosterone in the male rat fetus. This would soon be followed by a more profound and prolonged decrease in this hormone. What is interesting is that male baby rats, later in adulthood, will show demasculation (sexually mount other males). These could be "gay rats" as opposed to "straight rats." This is further evidence of this "window" of hormonal timing during fetal development that might dictate sexual direction or preference. What is also interesting, although perhaps less relevant, is the fact that the female offspring, when they become mothers, also have an increased tendency to have male rats that will also show demasculation (unmalelike) behavior. However, current evidence indicates this offspring finding is not related to humans.

Structural Difference in the Male Homosexual Brain

It was previously believed that the preoptic nucleus appears the same between the sexes in humans. This older belief is changing. In the human brain, this area is part of the hypothalamus and is located deep in the center of the brain. A cluster of cells in this area, known as the interstitial nuclei, is known to control sexual behavior.

Simon LeVay, of the Salk Institute in San Diego, reported in August 1991 that an area the size of a grain of sand in the hypothalamus was markedly different between heterosexual males and homosexual males. This study was based on the results of comparing (microscopically) the brains from nineteen homosexual males and sixteen heterosexual males; all died from AIDS. This group of cells has only a few thousand neurons and is referred to as INAH3 by scientists. What is more interesting is that the study also consisted of six heterosexual female brains (who also died of AIDS). When all three groups were compared, it was discovered that the cluster of cells was significantly smaller (less than half) in the male homosexual and female heterosexual group as compared to the male heterosexual group. Thus the homosexual male brains were similar to the heterosexual female brains in appearance. This further suggests that the male homosexual brain is more female than male and its sexual orientation starts in utero.

Some scientists argue that sexual behavior might somehow influence this structural difference. However, when we compare this result with the previously described rat experiments it becomes clear that the cause of homosexuality is structured deep in the brain and originates in utero.

The "Twins Study"

A few months after the LeVay study was published, psychologist Michael Bailey (Northwest University) and psychiatrist Richard Pillard (Boston University) reported the results of their "twins study," December 1991. They found that of fifty-six identical twins, 52 percent were gay. Now of the fraternal (not identical) twins, 22 percent (lesser but still significant) were gay. When considering the fact that gays comprise only 7 to 10 percent of the general population, this study suggests a genetic origin for homosexuality. It is also interesting to note the correlation between how identical twins are and the tendency toward homosexuality. This further suggests a genetic origin to the homosexual, meaning its origin may not only be intrauterine but in the DNA itself.

The Lemming

When we talk of stress in pregnancy and the possible relationship with homosexual offspring, someone with a background in psychology or sociology will often bring up "the lemming theory" of homosexuality. The lemming is a small rodent related to a class called voles, which have small eyes and ears and short tails, such as a muskrat. The animal is fairly common to North America and the Scandinavian mountains. Now these animals do a most peculiar thing that has been of much discussion among psychologists for many years. They live together and slowly grow in population. At some point the population density becomes very high and something seems to snap in the lemming brains. No, they do not become homosexuals; rather, they migrate together by the thousands over great distances, devouring crops and creating much damage, only to drown in the sea. Often they just jump off cliffs into the sea. This appears to be an inborn (hardwired) system of population control, which may in the long run be important for the survival of the species.

Often we will hear the argument that homosexuality may be a form of population control in humans as well, seeing that men making love to men is not likely to foster childbirth; that the environment feeds back on the human biosphere and through this mechanism of in utero stress helps to slow growth of the population. According to this theory, homosexuals are not quirks of the biological machinery but rather are part of the system, perhaps even an important part. Now we are getting into the environmental/congenital argument and, as you can see, it is not a simple one. However, in my view this Malthusian theory, with its application to homosexuality, certainly requires much more data. At this time it is simply food for the intellectual appetite.

Why Are Most Homosexuals Male?

A common question, I often get asked is the following: "But, Doctor, I thought that homosexuals were homosexual because of a heightened sexual appetite, and the fact that most homosexuals are male is because males are generally more preoccupied with sex?"

First of all, I'm not sure that maleness is a prerequisite for sexual preoccupation. Next time you go the magazine counter at the grocery, look at the magazines and books for women. We are all sexual creatures. Females prefer the left hemisphere (verbal/temporal) for the root to sexual fun. Thus you will find many sexual stories. The males prefer the right hemisphere (visual/spatial) and thus the centerfold. Business has exploited this for many years and profited.

Now to answer the question. Ever since Alfred C. Kinsey reported, just after World War II, on the incidence of homosexuality, others have confirmed his original report, suggesting a significantly higher incidence of male homosexuality. However, to discuss this, I will first need to talk about embryology. Embryology is a field of science that deals with the study of the origin and development of an individual organism. In the case of medicine, it would be the study of the developing human organism from conception to birth. When one studies this field, one quickly concludes that the main fundamental substrate (template) of the mammal is the female. During the course of development, if some genetic matter is missing, there appears to be a rule that dictates the resultant mammal will be female; that something extra is required to make maleness and if that something is not quite complete, the effort is discarded, which results in a return to femaleness. I will explain.

It is known that every nucleus of each of our cells is composed of forty-six chromosomes arranged as twenty-three pairs (twenty-three from the mother and twenty-three from the father). These chromosomes are a complex nucleic acid that are arranged in coded form. All the information needed to form a living human individual is coded on these twenty-three pairs. One of these pairs determines the sex of the individual and these are called the sex chromosomes. The female combination is called XX and the male XY because of their shape when viewed through a microscope. One chromosome is contributed by each parent. The X is always from the mother. An X or Y is from the father. This explains why the male sperm determines the sex of the organism. In the female XX one X is for the most part inactivated. This is called the Lyon hypothesis.

There is an interesting disorder called Turner's syndrome. This occurs when there is only one X chromosome (XO). The incidence occurs in about one out of three thousand live females. Physical characteristics include webbing of the neck, short stature, wide chest, and failure of breast development. However, all are female.

This is an example of the female physical structure being the fundamental mammalian pattern. When genetic material is missing (in this case the Y chromosome), the resultant is female. Also, there is no OY syndrome for it is not compatible with life while XXX will often be normal, physically and mentally, and apparently have normal offspring. This is why scientists believe that the main human mammalian substrate is the female. Something extra, biochemically, is needed for the male body, and if something goes wrong during development, the course is abandoned and a new biochemical course is charted for femaleness.

This being true in physical (body) characteristics, why could this not be true in the brain as well? After all, the brain is part of one's physical make up. Why should it not also be subjected to the same rules of embryology; that during fetal brain development, if something is missing, the same rule of embryology would apply, resulting in a brain with a female sexual direction. I have called this Weisher's hypothesis of neuroembryology. However, I may not have been the first to cultivate this idea. Such a theory would explain the reason for the higher incidence of male homosexuality for many changes could, theoretically, have the same end result of cerebral female sexualism. Thus, there would be more female brains as "end products" than male brains, regardless of the sex of the body.

Are Male Homosexuals Simply Female Brains in Male Bodies?

Are male homosexuals, then, simply female brains in male bodies? Or is it more complex? Is there a difference between a male homosexual brain and a normal (heterosexual) female brain?

Since Hooker's report in the *Journal of Protective Techniques* (1957), psychologists like him have been trying to dispel the myth that all male homosexuals are stereotypic and display typical female mannerisms. I have also found this myth to be untrue based on my own observations of gay males. Here are some relevant observations:

1. I noted that stationery used for letter writing often would have a background picture of a nude male figure. This is not typical female behavior. This would indicate a typically male (right hemisphere) orientation to sex.

2. There were some female mannerisms. Fairly often during emotional periods of conversation, there would be a lot of head movement and hand movement (hand to chest) more typical of the female. This might indicate a cerebral hardwired (congenital) feminine component).

3. Histories taken of patients by male homosexual doctors and residents were often of excellent quality in regard to detail, handwriting, grammar, and chronological order. This was also typical of many of the female doctors I have known and would be consistent with higher skilled left hemisphere function (verbal/temporal) typical of the female. For example, surgical residents and doctors would often write very cryptic histories typical of a brain that might be lateralized to the right (visual/spatial).

4. Some were very feminine in mannerisms and speech. I do not feel this was acquired or purely cosmetic in nature. For

example, the grammar and pronunciation were accurate and careful. Males are often less careful in selection of words or speech pattern, possibly because most, being right hemisphere in skill orientation, have less developed verbal skills than their female counterparts.

5. Often they were very much concerned with feelings; this would include their own feelings as well as concern for others to a point one might consider "nurturing." There appeared a genuine concern for those unjustly accused or prejudiced against. Could this be socially acquired because of their own fear of prejudice or a deeper, hardwired sensitivity typically female?

6. Many had the typically male ability to differentiate love making from love. No doubt, this contributed to increased promiscuity.

These are simple observations that raised questions regarding the makeup of homosexuality. Male homosexuality may be a complex combination of right hemisphere and left hemisphere skills and not simply a female brain in a male body.

Why the Increase in Promiscuity?

Promiscuity in male homosexuality is a well-studied observation. There may be several reasons for higher promiscuity in this group:

1. The male hormone testosterone, if given chronically (long term), can induce more aggressive behavior, which also includes sexual aggression. This hormone does not change sexual preference in the adult. The male homosexual is exposed to the exact same hormone and at the same levels in the blood as any other male.

2. The intersexual bonding forces are between two more sexually aggressive animals. Therefore, sexual hyperactivity and perhaps promiscuity are further stimulated.

3. Autovisual stimulation could also be a part. Could the constant exposure to male sex hormone allow the brain to acquire a more right-hemispheric (visual/spatial) appreciation of the nude male body, resulting in a "higher skilled" visual-sexual

recognition? Autostimulation (looking in the mirror in the nude) would be unavoidable, and would naturally result in a higher sexual preoccupation.

4. A psychological reason might be fear of unavailability as there are fewer gays than heterosexuals to pick from. This might be a sort of supply-and-demand effect.

5. Another is the typical male ability of being able to separate sex from emotional involvement, as described by Saghir and Robins (1969). Thus, sex becomes recreational as well as romance and is viewed with the same degree of expectations from each member.

It would seem that sexual hyperactivity would be almost inescapable in the male homosexual. Notice I am *not* saying promiscuity is inescapable. In my experience with the gay population, at times of death and dying there was still the same love bond and caring between couples as one would expect from the heterosexual. They are just as capable of forming that love bond as anyone else.

This discussion of homosexuality should not be confused with homosexual behavior as a result of incarceration. The "jail bird" homosexual relation is nothing more than a substitution for normal heterosexual behavior and stops after incarceration.

Summary

Although there is considerable evidence that homosexuality may be determined in utero, it might be more complex than just a female brain in a male body. There may be a unique combination of left and right hemisphere skills that influences behavior. There should be little fear, for homosexuality is not a disease. It is not spread either from an airborne, social, or even genetic (dominant/recessive) causes. Perhaps with a little education, there will be a reduction in fear and prejudice. With a little education, we might come to realize that homosexuals are, indeed, different but equal.

4. The Concept of Soul/Individual

I remember as a child looking in the mirror and noticing how the pupil of my eye would constrict with exposure to light. Mechanisms of all kinds fascinated me, and I was often preoccupied by anything mechanical. This response of the pupil to light caught my attention because here was an example of the concept "man the machine." I used to stand back from the mirror and wonder if I were *just* a machine, as evolutionists would lead us to believe, or if there were something more, something of "spiritual" significance, something that we could never understand because the added entity might be beyond the capability of the intellect. You see, the pupil is physical evidence of an analog computer in the subconscious or brain stem area that calculates the amount of light entering the retina (back of the eye). It then "feeds back" to the pupil and therefore regulates the amount of light so as not to overwhelm the visual apparatus. Try it the next time you look in the mirror and see how you feel.

We are, essentially, our brains. It is well known in medicine that there is a hierarchy of survival in the human biosystem that mandates the brain to be given the highest chance of survival in instances of extreme trauma. All other systems are there to serve the brain. The lungs provide it with oxygen and the heart pumps the blood provided by the bone marrow to this brain, the highest consumer of energy. Just think the next time you are physically threatened on how you positioned yourself to protect your head. This reflex was not acquired but was built into your physical defense mechanisms. Notice how the brain is specially protected by that cranial vault we call the skull. However, man did not always look upon the brain as the origin of thought. The ancient Hebrews saw the heart as the intellectual center of the body, while the Arabs thought it was the liver. Today we know that without the brain we cease to exist.

However, the next question that may come to mind is, Where is the "individual" in all this? What is it that separates you as a unique "person" unlike anything in existence now or perhaps in the future as well? Just as we can divide the brain into functional areas, as described in the first chapter, is there an area in the brain that we can call the individual person or are we simply a compilation of the whole neurocircuitry? No one can doubt that this is the highest of man's intellectual and philosophical challenges. Is the "individual" hardwired into the brain from conception (hardware) or developed as a response to the environment (software)?

The hardware/software argument is perhaps the oldest in neuropsychiatry. Modern psychiatry and the psychogenic theory began with Austrian physician Fredrich Mesmer (1733–1815). Mesmer was fascinated with the problem of hysteria. Hysteria is when a patient complains of an apparently physical disability that they are absolutely convinced they have but for which there is no organic (physical) problem ("It's all in your mind," as some would say).

Mesmer proposed a bizarre theory that an alteration of "magnetic fluids" about the body was the cause of hysteria. He developed an elaborate and theatrical way of "adjusting" these fields (dimming the lights wearing a cape and touching patients with a magic wand). Thus, the term *mesmerized* and *animal magnetism* came into being. It seemed that many patients were actually helped by this treatment. However, Mesmer was later barred from practice.

After this, Ambrose Liebault (1823–1904) and Hippolyte Bernheim (1840–1919) discovered the Nancy School (Nancy is a city in France). This school practiced the art of hypnosis with some apparent success. However, one of the most famous names in neurology of that time, Jean-Martin Charcot (1825–93) from Paris, strongly debated its validity, feeling that most if not all causes were organic (hardware). The debates were heated, and perhaps this was the first time in history that the hardware/software concept was seriously debated.

Later, around 1950, the British mathematician Alan Turing presented an unusual case scenario that he felt favored the concept of man the machine. Computers were beginning to be commonly known, although they were still in a primitive state of development. He hypothetically described a woman in one room and a computer in the other. Questions and answers were presented by way of a keyboard by a third party. Turing argued that if the computer could be programmed to respond to questions as convincingly as a human in another room (where it would be difficult for the administrator to distinguish human from machine), could it therefore be determined that the computer might possess, in some small way, the equivalent of human intelligence? Later, as artificial intelligence (AI) grew as a field with the development of computers, this scenario became known as the Turing test.

Could a machine, no matter how sophisticated, be able to possess the equivalent of human intelligence? Could a machine be the equivalent of human life? In other words, is what separates man from machine simply a degree of sophistication or is there something more? Are we self-programmable machines with our environment feeding back and affecting our development through the senses, or is there more? These are deep neurophilosophical questions that become more intriguing as computer hardware and software become more advanced. I should hope no one would fall in love with a machine, although psychiatrists in the next century may be dealing with just such a problem, when computers capable of both hearing and talking become commonplace.

Some might call this the religion of neurophilosophy. On one side we have spiritualists, who believe that the human mind cannot be reduced to "simple" laws of physics; they feel the individual lies in the spirit that has control of the brain. Mechanists, however, feel that all human functions can be reduced to physical laws and anything more is simply just hocus pocus or swamp gas. No one knows the answer, but seeing that we live in a physical world, let's examine the various physical views.

Me the Machine

One may ask, "Am I a function of my environment or is my

environment a function of me?" Is the area of the brain that determines the individual predetermined at birth? Or is it gradually shaped by the stimulus of environment just after birth? It is easy to see how one can say, "I have a liver," or, "I have a spleen." And seeing that we are our brains, one can also say, "It is me, my brain, that has these things." However, our brain is also composed of parts just as our body is. These include the thalamus, the cerebellum, the cortex, the basal ganglia, etc. One can also say, "I have a thalamus," or, "I have a basal ganglia." Then the next question arises. "Where is the 'me' that has these parts?"

Cerebral lateralization describes a brain divided into various areas of skill. Where is that portion of the brain that can no longer be divided, the true possessor of all other organ systems? The word *individual* comes from the Middle Latin word *individualis* and the Latin *individuus*. Both of these words mean not divisible or not separable. Thus, we are speaking of an area of the brain that is unique to the person; it is indivisible, not a difference from other species but from others within the same species. These are deep subjects and involve all kinds of ramifications. During various parties at Georgetown, my colleagues and I would gather together after a cognac or champagne and have fun with this puzzle. I would like to share some thoughts with you. There is as, of yet, no clear answer, but I would like to give you some food for thought.

During my residency at Georgetown, I saw an impressive patient from Iraq. Half of his brain had been blown away by shrapnel during the Iran/Iraq war, and he was sent to us to see what could be done. If *self* is the collective function of the whole brain, would this person be less an individual? I think not, but then where is that one single area in the brain where it could be said, This is the person, or This is "self" or "individual" (that this one area of the brain determines the unique person that you are)? Some would argue that it is not a "thing" but a spiritual entity that is bestowed us during conception, never to be understood. Who knows the answer to this "entrée for the intellect?"

The Brain Switch

As discussed, we are our brains. Man has been fascinated with the concept of a "brain switch" for many years. Scenes of Boris Karloff on the silver screen, playing the mad scientist, about to put the brain of a young girl into a gorilla, come to mind. However, did you know that similar things on a smaller scale have been done in the laboratory? Nicole Le Douarin of the Institut d'Embryologie Cellulaire in Nogent-sur-Marne, France, reported in *Science* (vol. 241, 1988) that they have changed the behavior of one bird species (a chick) by the transplantation of part of the brain of another species (a quail). They have observed that the recipient chicks sound a lot like quails.

These transplantations were performed during embryonic development. They removed part of the neuroepithelium from a young quail embryo and, after removing the corresponding brain region of the chick embryo, placed the quail neurotissue into the chick brain. This is called a quail-chick chimera (*chimera* comes from the Greek, *chiasmos,* meaning to place crosswise). These observations were made during the first week of life because after the second week, the birds began to immunologically reject the grafted brain tissue and die. Were these chicks stripped of their "individuality," thus having the quail brain in "ownership" of all chick-organ systems? Or was there another individual formed? We just don't know, but in order to postulate, we need to look at some of the theories regarding the cerebral location of the individual.

The Cortex Theory

There are many who believe that what separates the human mind from other animals is the ability to make tools. We are able to create a device that is helpful in solving a problem (a hammer, screwdriver, spear, bow and arrow, etc.). Some people believe that only humans have a "soul" and that all other creatures are somehow different and simply react predictably to environmental changes. Dogs chase cats, cats chase mice, etc. Thus, it is felt, they lose their individuality because they lack an advanced cerebral cortex, such as is found in the human brain.

What happens when a cat loses his/her cerebral cortex? The process of removing the cortex of the brain in an animal is referred to as high decerebration. If the cortex of the cat is removed along with some of the rhinencephalon (the portion of the cortex responsible for sense of smell), the animal will appear unchanged. He will purr just like a regular cat. However, at the most innocuous stimulation, the cat will show tremendous rage grossly out of proportion to the stimulus. This rage is often referred to as "sham rage" because it is felt that the animal is not able to feel or direct anger. This may be debatable. However, we can assume the animal is no longer able to reason about the type or degree of stimulation so the response is stereotypic (the same) and extreme. The cat no longer has a unique personality. Could this mean that the cortex is the source of the individual? If so, then why does the decorticated cat still purr like any other cat and is still able to exist? Does this fact negate the theory? This is the cognitive theory of the soul.

The Limbic Theory

Some feel that because man is an emotional animal, the individual must lie in the areas of the brain that control emotion (the limbus).

This deep and centrally situated gray matter has a margin about the corpus callosum and surrounds the midbrain (the upper part of the brain stem). The word *limbic* comes from the Latin *limbus,* which means margin. This word is used because there is a margin of the cortex that folds inward at the temporal lobe, forming the limbic area. Phylogenetically, this is one of the oldest areas of the human brain. This part is believed to be related to emotion and behavior. It is also related to smell and in lower animals is called *rhinencephalon,* which comes from the Greek *rhis,* for nose. It is intimately involved with the formation of new memory or what I call "data entry zones." It appears that information to be stored passes through this area on the way to

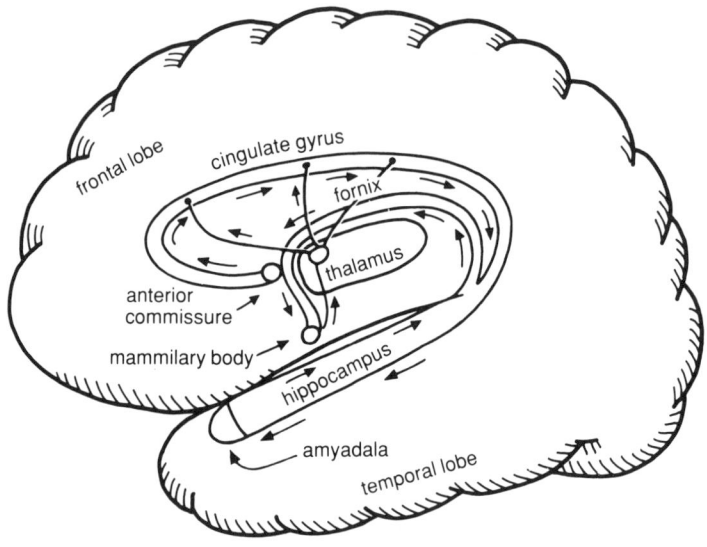

Fig. 15. Circuit of Papez limbus.

memory banks in the cortex. What happens to the animal when parts of this area are ablated (cut out)?

The very tip of the temporal lobe in lower animals, such as the cat, is called the piriform area. If experimental lesions are produced here on both sides (bilaterally), the animal's behavior is most interesting. The cat will become very tame and not have fear, not even of natural enemies, such as dogs. They will eat anything, including wood such as match sticks (hyperorality), and the males, if the male sex hormone (androgen) is still present in the blood, will be hypersexual. Perhaps one can conclude from these experiments that sex is an appetite. In the laboratory animal, this is often called the Kluver Bucy syndrome.

The full behavior pattern of these lesions has traditionally been felt not to occur in the human. However, I have seen several patients who have suffered bitemporal strokes that resulted in similar lesions as in the animal studies. Little is changed in the human, except for the fact that he is virtually unable to form new memory. For example, you could enter his room, introduce yourself, and have pleasant conversation for several minutes. But if you return one or two minutes later, the patient will have no recollection of the previous visit. Often, however, the patient will be able to recall old memory and talk about his youth or occupation, but he will be unable to form new memory. If the neurological exam is rushed, this profound defect in cerebral function can be missed and sometimes is by careless medical students.

The Kluver Bucy syndrome was previously believed to occur only in animals. However, recently, some have reported this syndrome in the human brain. V. Olivera and J. M. Ferro report, in the *Journal of Neurology* (1989), a case of a forty-eight-year-old woman with multiple brain infarcts (lesions) as a result of lupus vasculopathy. This is a vascular disease that sometimes involves the vessels of the brain. This patient suffered extensive damage to both temporal lobes. She developed a complex behavioral pattern, which included hyperorality, global aphasia, and hypersexuality.

Do these bizarre behavioral changes shed any light on the location of the soul (individual) in the brain? These two areas of the brain represent two main concepts or views on the course of the individual from opposite points (emotional versus intellectual). The cortex theory states that the "thing" that separates us from one another is the degree of intellectual ability. The limbic view looks at man, the emotional animal, and suggests that the emotional response is what differentiates us and that the cortex is an added intellectual benefit.

I have believed that the study of embryology (the study of the developing human) would be helpful in generating an appreciation of the brain and may shed light on the location of self. For at some time during development, self is born.

The Embryology of Self

Around the fifth week of fetal life, the human brain is nothing more than a membrane balloon called the forebrain or prosencephalon. *Pros* comes from the Greek *proso,* meaning before. In the sixth week, the top portion divides into two bubbles, and these become the two halves of the cerebral cortex. This is now called the telencephalon from the Greek *telos* (end). The

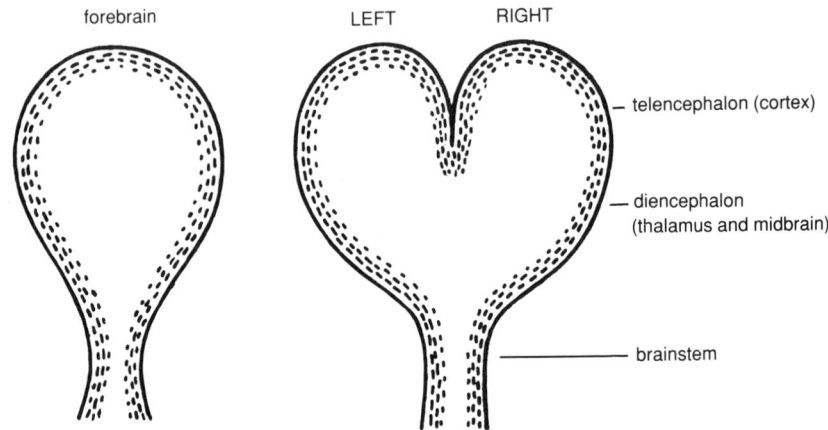

Fig. 16. Fetal brain development.

lower portion of the forebrain remains and is now called the diencephalon from the Greek *di* (through or between) because this part lies between the cerebrum and the brain stem.

The formation of these parts is akin to a balloon sculpture forming an animal. Several bubbles arise from one balloon. After all this is done, on the seventh week, something very interesting begins to happen. The hemispheric wall actually begins to become thick. On careful observation, it is seen that the increase in thickness is because of a rapidly increased growth of cells that comes in waves, originating from the innermost portion of the membrane and traveling outward.

As noted in previous chapters, we often speak of hardwiring in the brain when thinking of behavior that is based on heredity or congenital development. It is here and now that hardwiring is being formed: our male and/or female characteristics, the left hemisphere or right hemisphere orientation, as well as homosexuality versus heterosexuality.

The innermost membrane layer, where these cells originate, is now called the germinal mantle. These cells, which rapidly divide and migrate outward, are called *neuroblasts,* from the Greek *blasto* (germ). They migrate in waves. The first to migrate are the astrocytes. This makes sense because astrocyte cells, not being actual neurons, are supportive in function. They help supply the neuron with glucose for nutrients and they take care of waste products produced by the neuron. Another name for astrocyte is from the Greek, *glio*, meaning glue. This is because early investigators, while looking through microscopes, thought that the primary function of this cell was mechanically supportive like a cast (glue) to physically support the neurons (much like a conduit supports electrical cable). This way neurons, which later migrate outward from the germinal mantle in waves, will have the necessary supportive environment to live.

This process of cellular-wave migration goes on for the next four months, and by the sixth month of fetal life, all the six basic cell layers of the cerebral cortex are formed. It is interesting to note that the first neurons to mature are the larger pyramidal cells. These are named for their shape and are involved with voluntary muscle control. The fetal brain at three months is smooth, and there are no sulci or gyri. Later, at six months, the central (rolandic) sulcus begins to form and along with it the pre- and post-central gyrus (motor strip and sensory strip, respectively), as described in the first chapter. By eight months (thirty-two weeks), a total gyril pattern is achieved and the brain looks much like a human adult, although smaller. By the age of six years, the human brain will obtain 90 percent of the weight of a mature adult brain.

Glio/Cell Index

As the cerebral cortex matures, there is a gradual decrease in neuron cell density and an increase in the glio/neuron ratio (glio/cell index). This helps support the view that we are given only so many neurons, about 10 billion, and no matter how much we "exercise" our brains, it will not develop more. However, one could speculate that as we use various areas of our brain, they become more "mature" and thus have a higher metabolic demand. A more active area of the brain requires more supportive cells (glio) to handle the metabolic demands, glio/cell index. Glio cells are responsible for nourishment and other supportive roles to the neuron. It was discovered that an area of the brain, Brodmann 39

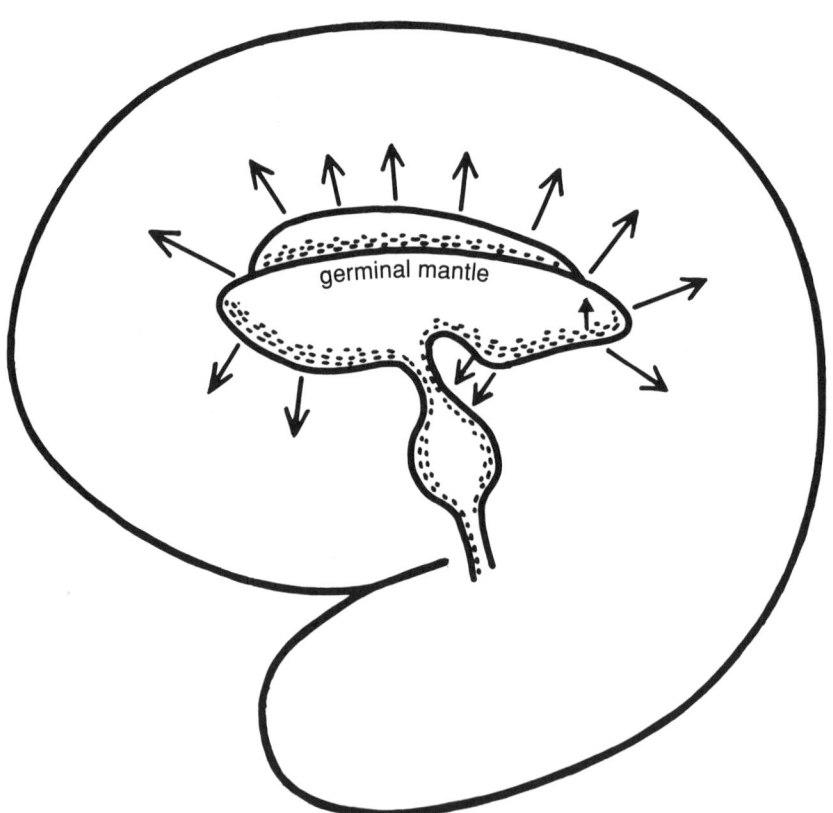

Fig. 17. Cell migration in the developing brain.

(posterior temporal), has a higher glio/cell index in persons with professions or hobbies that require much intellectual exercise. Albert Einstein had a high index in this area. Perhaps you are also affecting your glio/cell index by reading this book? Some psychiatrists use this concept to illustrate environmental influence on cerebral hardware and therefore behavior as well.

It might be during this period of neuronal-cell migration that the individual is formed. Is it the astronomical permutations and combinations (variability) that set us apart from one another and make us truly unique with no other like ourselves? Is this the key to self? Or is there more? Is the cortex or limbus simply an intellectual servant of another area of the brain that is the individual? Are there other ideas regarding the formation and placement of self?

The Genetics of the Self

When we discuss the concept of soul, neuronal embryology, and neuroblast migration, DNA must come to mind. As you know, DNA is the coded template of the human organism and is unique, to some degree, for each of us. One can, therefore, presume that the self or soul is also coded there. One can assert that we are all different but equal. We are different, yet the same. The criteria for individuality is not one of superior versus inferior but rather a difference among equals within a given species. It appears that this difference is recorded on the DNA.

DNA (deoxyribose nucleic acid) is the template or blueprint of the human organism, including the brain. It is present in the nucleus of all cells in the body. Even the squamous (flat) cells secreted by saliva have the complete package of DNA to form the individual. DNA is composed of a double helix. Two "threads" of this material are right-handedly coiled (double helix) and then coiled again.

Located along this double thread are various combinations of four organic bases—adenine, guanine, thymine, and cytosine—used for coding. Needless to say, the various combinations are astronomical. For example, the various possible sequence of arrangements of a DNA of only fifty bases is four plus fifty zeros. Obviously, the human mind is not capable of grasping the various combinations of a normal DNA of millions of base pairs. You can quickly appreciate the astronomical amount of instructions needed to "manufacture" the simplest organism, let alone an entire human being.

Protein is the main component of the mammalian organism besides water. Protein composes bone marrow, skin, blood cells, muscles, neurons, etc. The instructions for the formation of these various proteins come from the areas of the DNA called the genes. There are about 50,000 to 100,000 estimated genes in human DNA. V. A. McKusic (1988) reported, in the "Mendelian Inheritance in Man," that only 4,600 are known at present. Of these, about 1,500 have been "mapped" to one of the twenty-three pairs of human chromosomes. Of these, only 600 have been sequenced (order of base pairs). There has been talk of mapping and sequencing the entire human genome. The Committee on Mapping and Sequencing the Human Genome of the National Research Council of the National Academy of Sciences reports that the estimated cost to do this would be approximately $200 million a year. The time it would take to do this astronomical achievement is estimated at fifteen years, making the total cost almost $3 billion. There are also an estimated 3 billion base pairs in the genome of the human. If we were to type this out in a library and, assuming one base pair per character, the size would be equivalent to thirteen sets of the Encyclopedia Britannica. That's a lot of print.

Now that I have totally "mesmerized" you with the astronomical amount of instructions needed to form the human body, would you believe that the estimated fraction of DNA for genes is only 5 percent? As mentioned previously, genes compose the necessary instruction for the formation of protein, which is the main composition of the body. Now, this is the $64,000 question: What is the other 95 percent of the DNA instructed to create? It must be for something. Could it be that here lie the genetics of the individual or soul? Is it possible that this major portion of human DNA is for genetic instructions necessary to form you personally? That within this astronomic myriad of coded genetic material lie the schematics for the development of the human brain and therefore the individual (you)? If this is true and you were to die and your full genetic information left behind in a test tube, you could be reconstructed (resurrected). Needless to say, you would be unable to recall events from your previous life or even be aware that you existed prior to your birth. The reason for this, of course, is that memory is software much like a computer. If the computer is turned off and then on, all memory in the CPU (central processing unit) is gone. This provocative neurophilosophical discussion can blossom into many possible case scenarios.

What about Twins?

The next logical question is what of monozygotic (identical) twins? They have extremely similar, if not exact, genetic (DNA) material. We can't have two persons living simultaneously and both the same individual; that would defeat the very definition of the word and concept. It is no secret that monozygotic twins often marry similar mates, engage in similar careers and hobbies, have similar personalities, and live similar life-styles. (Thus giving credence for our lateralization theory of behavior following skills.) Needless to say, there must be considerable genetic influence. Nonetheless, they are separate individuals. If we cling to our "mechanistic" as opposed to "spiritual" theory, we then must accept two explanations for this: they are the "dynamic" and the "environmental" neurophilosophical theory of the development of the soul.

Dynamic Theory of Self-Development

This would explain the separate individuals of the monozygotic twins on the premise that the rest of the DNA during development is a relatively dynamic/changing structure. Thus both monozygotic twins could never be the same "individual." There will always be a difference in the hardwiring of the brain because the codes instructing the wiring are always changing, to some extent, during development.

Environmental Theory of Self-Development

This theory states that the DNA is relatively fixed during development and does not change. Different hardwiring of the cerebral cortex or other areas of the brain necessary to form the individual is dependent on the relative position of the developing brain in space and gravity. Thus, no two objects duplicate in the same pattern of space and time. Therefore, there is always a difference in the hardwiring; hence we have two separate individuals. Those interested in astrology might find this theory acceptable as one could postulate that astronomical bodies and sunspots may have an effect on the formation of the individual and therefore behavior later in adult life.

Spiritual Theory

We must also consider the spiritual side of things. This theory states that man could never understand the concept of self or soul because its content is beyond the material world. Man, being a material animal, cannot fully comprehend the nonmaterial (spiritual) side of life; that the living human organism is assigned a soul at conception and thus becomes "whole."

Personally, I do not feel that any of these theories is against God or the concept of God. Some may feel differently. However, not being one to place limits on the Lord, I will not discredit any for life itself is a sort of miracle. A true scientist is always open minded. However, I must say that as we learn more and appreciate the tremendous complexity of the human brain and mind, this theory becomes more convincing.

The Thalamic Theory

The thalamus is a small gray-matter structure that lies at the center of the human brain. The name comes from the Greek, *thalamos,* which means chamber. Thus it is the inner chamber of the brain. One of its major functions is to act as a switchboard for incoming information (signals), such as pain, temperature, vision, hearing, etc. It then relays this data outward to the cerebral cortex. All incoming information must first be "processed" in the thalamus before it is brought to consciousness in the cortex. For that reason it has been called the "gateway to consciousness." It is also important with the formation of new memory (data storage). It is intimately involved with maintaining a state of wakefulness and thus is part of the reticular-activating system. It also seems to act as a modulator of motor (muscle) control.

It is an amazing little structure that seems to have influence over all aspects of animal brain function. Its close proximity with the limbic (emotional) structures suggests that the thalamus might be involved with the emotional response to vision, hearing, feeling, etc. We humans are emotional animals. Thus, when John and Jane both see a beautiful woman disrobe, there are two different patterns of signal processing going on. The emotional and the cognitive processing of visual data, as seen in the picture; visual data from the eye enters the thalamus at the lateral geniculate body, gives off signals to the limbic system, then goes on the occipital lobe in the back of the brain. Thus we have the thalamic (emotional and memory) processing of visual information and the occipital (cognitive) processing of data. The male receives a different emotional response to the same visual stimuli compared to the female. Jane knows, via the cortical cognitive processing of the information, that the stripper is stripping just as John does, but she is not sexually or emotionally stimulated. However, the thalamic connections with the limbic structures might be different in John; thus, he is sexually stimulated whereas Jane is not. Perhaps God placed the thalamus in the center and had all sensory input first pass through this "signal processor" because he wanted us to be both emotional and intellectual. Its placement was one of simple circuit design.

As in all bodies of the brain, the thalamus is divided into a left side and a right side. Both, morphologically, are the same and are composed of the same nuclei, both in number and location when compared side to side. Thus, the left side is identical to the right.

It is known that stimulation of various areas of the thalamus will, in turn, activate specific areas of the cerebral cortex, while stimulating the thalamus in general will activate the entire cortex. It is for this reason that scientists believe man can only concentrate on one thing at a time. Neurologists have claimed for many years that this body is responsible for the generation of dominant rhythm brain waves seen on the EEG (brain wave test). In the normal brain we see beta rhythm (thirteen cycles per second or more) in both frontal lobes, while the posterior are alpha (eight to twelve cycles per second).

Why would the thalamus be involved in our discussion on the location of the individual or self? A good reason is the fact that this is the only structure in the brain that deals with the integration and processing of all senses (data entry/signal processing) of which its removal is incompatible with life. As described previously, an animal can live without the cerebral cortex: "high decerebration." However, if the entire thalamus (both sides) is removed, the animal dies. Could it be that the soul or individual is removed with the thalamus, causing death to the person?

Some may argue that because the animal does not expire when half (left or right) is destroyed and one cannot be "half an individual," we should look elsewhere for the location of self. However, those in favor of the thalamus as the seat of the individual point out that this organ is well duplicated. In other words, the left half is identical to the right (quite unlike the cortex as realized in cerebral lateralization studies) and that there are thus two thalami (right and left). Therefore self is preserved with the destruction of the left or right but not both. Perhaps the individual is somewhere in the thalamus. I believe this theory has merit and should be investigated further.

Summary

As of today, no one knows, with great certainty, the organ of individuality, be it cortex, subcortex, thalamus, or spirit. However, one can make an educated guess. No harm in that. This chapter was simply to help bestow an appreciation of the miracle we call life. Each one of us has in our possession the unique indivisible element of the individual (whatever and wherever it is) and thus there is a difference through equality.

5. Savant Syndromes and Genius

This chapter presents an interesting discussion of genius and savant syndromes by describing anecdotal histories that support cerebral lateralization as a likely mechanism for explaining unique and unusual abilities, such as "calendar brains," artistic skills, and musical talents.

In 1969, when I was studying engineering in college, I came across an article in a science journal by a colleague at MIT. This fellow had a novel idea about the mechanism of human creative thought. He had noticed that during times of stress, when attempting to solve very difficult technical problems, sometimes the solution would pop into his head with no apparent cause. At times he had no idea how he had arrived at the solution and so coined the phrase "subconscious logic." I have lost the article but I have never forgotten this novel concept. I do not recall the author's name but would like to credit him with his open mindedness, which is in the true spirit of science.

I recall a similar experience myself while working at Guarrett Air Research in 1977 on an extremely sophisticated piece of computer diagnostic hardware for the F-14 computers that suddenly failed. You can't just call a technician and say "fix it," as if it were a television or radio. This was a very large, complicated, and custom-designed piece of computer hardware. Only those intimately involved with its functions were able to eventually solve the problem. But if this computer monster was not repaired in time, it could have been a matter of national defense, for the F-14 was the primary strategic defense fighter aircraft for the U.S. Navy. No time was to be wasted.

It should be pointed out that no one person could be completely familiar with all of the technical aspects of this system. However, many of us had at least been exposed to all the circuit diagrams (schematics). During the course of about twenty-four to forty-eight hours of intense thought, I developed a most stunning sensation. For some reason I believed I knew the chip that was causing the problem. These particular chips were soldered to the board and thus were very difficult to interchange. No other chips on the entire system were changed. I could not say why I felt so strongly about my diagnosis, but I carefully extracted the chip and replaced it. Low and behold, the entire system became operational and I was a big hero. However, I must admit, I did not feel particularly bright, for this was simply subconscious logic at work.

Another event in my life that involved a brush with subconscious logic is a night sailing story. It was a dark, cool night in the fall of 1990. I was with a good friend sailing from Oxford, Maryland, on the Choptank River, going west. Our destination was Harrington Harbor across the Chesapeake Bay on the western shore. To save time, I planned to go through Knap Narrows in order to enter the Chesapeake and continue west. I had planned the course carefully, and although it was a beautiful night with a sky full of stars, I was tired and decided to rest below while a friend was at the helm, maintaining course. While I was relaxing, a fearful feeling came over me. Something was terribly wrong. I grabbed a high-powered flashlight, went topside, ran to the bow, and flashed the light to where we were heading. To my shock and horror we were running straight into a rocky breakwater shore just south of the entrance to Knap Narrows. I yelled, "Reverse engine!! Full speed astern!!!" My friend was perplexed but fortunately followed the orders and we were spared a dangerous and costly accident. What caused me to suddenly realize that something was wrong? Could there have been a "little navigator" in my subconscious that knew the chart and our location by knowing course, speed, and time? Were these events simply chance, or could they be a manifestation of an undeveloped skill often seen in the idiot savant? If so, is this skill available to each of us? Many have similar stories; the question is, can it be cultivated? Is there a little genius in us all?

The word *savant* is French and means to have knowledge. It comes from the word *savoir,* meaning to "know." The term *idiot savant,* although cruel sounding, simply means a person who is grossly mentally handicapped but is endowed with a special talent or skill. The following are some examples.

The Savant

Derrick

Derrick was a very pleasant little boy, born blind and with poor verbal skills. He lived in a special home near London, England, and had a most intriguing ability of interest to all neurophilosophers: he had never taken serious music lessons; however, it was discovered that he was able to play the piano extremely well. After further investigation, it was realized that Derrick was able to almost instantly reproduce melodies heard on the piano; so his skill was apparently more reproduction, as opposed to pure creativity—a marvel none the less. Derrick has given piano concerts, playing the great classics with the most amazing skill. How was he able to do this? Is there any insight into our own cerebral function or intellectual potential that can be gained by studying this interesting phenomenon?

Leslie Lemke

Perhaps a much more famous story is the one of Leslie Lemke, who had a verbal IQ in the retarded range. He was born prematurely in 1952. Leslie, too, suffered from blindness and he, too, had an amazing ability of instantly reproducing melodies on the piano. You might have seen his story on "60 Minutes," first aired in 1983. Although Leslie's adopted mother had a piano in the house and had played the keyboard with Leslie as a child, Leslie displayed no real talent at first. However, with practice he did improve, but he continued to display no true virtuosity. Then, late one night, his mother woke up to the sound of beautiful piano music. Thinking that the television had been left on, she got up, only to discover Leslie playing Tchaikovsky's Piano Concerto No.

1 to perfection. He had heard the music on "The Late Movie" just moments before. Leslie's music repertoire grew rapidly, including popular and classics, and he has given numerous concerts.

Having special skills does not always mean that one is grossly retarded. The following is a good example of a savant (pure) who is not mentally handicapped.

Ira

Ira was a very good friend of mine. He was always kind, helpful, and polite. However, he had very few close friends, for Ira was "different." This was nothing new to him. Ira knew he was different from others, being an albino with a congenital ocular nystagmus (jerking eye). However, he had a unique talent—a photographic memory. For example, Ira and I were in the same class in medical school. One day when the results of the finals were to be reported, Ira, myself, and ten others were all in a room off the main campus, wondering about our fate. The medical school had the custom of reporting grades on a bulletin board. The grades were listed by our six-digit student ID number so as to afford some privacy. We knew the grades were to be reported that day, but why should all of us go when only one was needed to get our grades? Sweet Ira was elected. He would point his finger at us and say, "Give me your number." A few moments later, he would say, "I got it," and go to the next person. He did with all ten students. Nothing was written down. Ira went to the campus, checked the chart, recorded to memory the student number and the corresponding grade for all ten students, and returned. He was 100 percent accurate in his report of our grades.

Other medical students thought Ira was a freak, paying little attention. However, I had gone to medical school to be a neurologist, and in the true spirit of neurophilosophy, I asked many questions about Ira's skill. He was very open and friendly. I found that he was also able to perform fairly complicated arithmetic in his head. He did not date women, nor did he have any desire to, although he was not gay. His only interest was statistics, baseball, medicine, philosophy, and politics. He had no domestic skills. If no one cooked for him, Ira would live on Coca-Cola, which his refrigerator was always full of. How was he able to perform these interesting skills?

After all my questions, I discovered that the common denominator seemed to be that Ira's mind worked like a chalkboard, a sort of fluorescent screen. He was able to see in his mind. You might say his introspective vision was twenty-twenty. He could draw all the figures in his mind and do with them as he chose. Could this indicate that his skills were of the right hemisphere (visual/spatial)? It's also interesting to note that his apparently decreased domestic skills might have been a reflection of a relatively hypofunctioning left (practical) hemisphere, as described in chapter 2. Could it be that Ira was lateralized to the right, with some sacrifice of the left hemisphere, the total amount of cerebral skills thus being equal? Could this be an example of a tendency for the cerebral hemispheres to be conserved, a relative law of conservation between the hemispheres that makes us all different but equal?

The Genius of Earlswood

This is an old, well-known story. Earlswood was an institution for the mentally retarded in England. This story was reported by F. Sano in the *Journal of Mental Science* in 1918. The individual described had marked retardation in verbal and reading skills. His name was James Pullen, and he was considered to be deaf and dumb. On first impression, one would consider him the usual mentally retarded person. However, James had extremely proficient artistic skills; he was able to perform the most beautiful works of sculpture. Reportedly, until the age of seven, the one word consistently verbalized by him was "muv-ver," perhaps meaning mother. After he entered the institute of Earlswood at the age of fifteen, it was discovered that James had an amazing talent for carving model ships. Interestingly, he had a brother at the same institute. He, too, was deaf and dumb and reportedly was gifted in carving skills. The parents were normal and had no history of mental illness.

How good was James? He was so good that even the queen of England expressed an interest in his work. Because of this, James became famous and was given two workshops at the institute in which to express his expert craftsmanship. At the age of thirty-five, James began creating his greatest masterpiece, "The *Great Eastern*," from which he received first prize at the Fisheries Exhibition in 1883. This model ship was approximately ten feet long and was of exacting detail, even including the furniture. There were over five thousand rivets and the project took about seven years. Yet James could barely talk and was utterly unable to take care of himself. It is interesting to note that he was fairly egocentric about his work. Most times James was rather quiet and worked alone. He did not easily accept the company of strangers and was at times given to bursts of temper.

Was this skill acquired or was it the result of aberrant cerebral development? Some experts of James's day felt that his skills were the result of sensory deprivation (being deaf and dumb) and were thus acquired. However, his degree of skill was so highly advanced that this appears unlikely.

James's brain was obviously overdeveloped somewhere. The question is where? We know he must have been underdeveloped in the left hemisphere for he had very poor verbal skills. We also know that sculpture requires a high degree of manual dexterity and that the frontal area of the cerebrum is dedicated to motor control. Could he then be overdeveloped in the frontal regions? After all, sculpture is a manual skill.

Well, the genius died at about the age of eighty years, and his brain underwent autopsy. James's cerebral cortex was then compared with controls (normal brains). It was discovered that James's frontal lobes were, in fact, small. Also, his temporal lobes were small. His corpus callosum was fairly large (fibers connecting the left and right hemispheres). However, the occipital lobes (the back of the brain used for sight) were massively developed when compared with the controls. Thus, it would appear that James, like Ira, would have been able to "see" the objects of art in his mind and reproduce them almost as if he were tracing them on paper. The skill was one of vision (seeing in the mind) not motor, as we might have suspected. As they say, "There is more than meets the eye." It is also interesting to note that James's brother reportedly had a similar skill, with difficulty in verbalizing as well. So perhaps genetics does play a part. It should be pointed out that, according to Eidelberg and Galaburda (1984), the area PEG of James's right hemisphere was found to be structurally similar to the visual cortices (occipital lobes) on microscope (see chapter 1). This would be consistent with the current view of a right hemisphere lateralization of skills in painting and sculpture. Could this then also be an indication of a relative conservation of function between the hemispheres, that maybe extreme hyperfunctionalism of one hemisphere is gained at the relative expense of the other side of the brain? Thus lateral (side to side) cerebral functioning may be conserved and would thus, as well, be different but equal.

Lateralization and the Idiot Savant

Why are people with such strange abilities in the midst of so much intellectual disability referred to as "idiot savants"? They are sadly called idiots because of their sometimes startling difficulties with ordinarily simple tasks, such as addition, subtraction, and verbal skills. This name was first given by Dr. J. Langdon Down, in 1887. He was the superintendent and chief psychiatrist at the Earlswood Institute. It is interesting to note that he was the one who first described Down syndrome (trisomy 21). However, in the case of savants, he made two very important observations that still hold today: (1) savants often have normal parents with no significant mental history, and (2) savants are almost always male. To this day there are relatively few known female savants. Could this be a clue that cerebral lateralization is involved in the development of the savant? Could this also be a clue to sexual differentiation in the normal adult human brain?

Another common presentation is the combination of blindness and highly developed music skills in the idiot savant. Is this a key to understanding a possible developing visual/spatial compensation and its relationship to music?

Idiots savants appear to be simply mentally retarded; what sets them apart is their amazing skill at one specific function. These focal areas of high skill are sometimes referred to as "islets of intelligence." There are three main common areas of cerebral superfunctioning that are of particular interest.

Common Talents in Savants

1. A talent frequently exhibited by savants is in the area of music. These individuals are able to reproduce heard music. They are not clearly known to create their own; rather they reproduce melodies on a particular instrument (most commonly the piano). Often they are able to reproduce the music only a few moments after hearing it for the first time. Also, they are not able to fully explain how they are able to do so. Sometimes the parents will say the patient one day just started playing the piano, with no lessons and making very few mistakes. How are they able to do this? Is there anything that we can learn about our own brain from these examples? However, in fact, there has usually been some prior exposure to the instrument.

2. Another common area of skill is the so-called calendar brain. These individuals are able to give the day of the week of any given date (month, day, year). For example, they might be asked what day of the week did September 20th, 1542, fall on. They will then answer Tuesday, often faster than a computer. Again, they are not capable of giving any explanation for their skill. They will simply say, "I just know." How are they able to do this? Are there any hidden neurophilosophical truths in the pattern of these skills?

3. Finally, we come to sculpture. These persons will mold with their hands and/or simple tools the most beautiful works of art in clay. Often the subject will be animals or some other known common object. The detail is often amazing, the muscular anatomy exact in all ways. Again, often they will not be able to adequately explain their skills. However, during an interview, I heard a very interesting explanation. The savant said, "I can see

it in my head." Does this give us any clue to cerebral functioning and right hemisphere (visual/spatial) skills?

It should also be noted that some savants may have extraordinary skills in verbal memory, such as the complete commitment to memory of a phonebook or play. However, often this capacity is only achieved if the data is read by the savant, indicating a visual/spatial presentation of data rather than verbal. Others may be able to instantly obtain the result of multiplying two three-digit numbers (lightning calculation).

There are some interesting common denominators involving these three categories of savants. First of all, they all seem to possess right hemisphere skills: the calendar brain with numerical skills, and the sculptor with his highly developed visual/spatial skills. Another is the profound decrease in their left hemisphere skills, which are more practical and deal with verbal communication. In addition, they are not usually able to take care of themselves, which would indicate a difficulty with practical left hemisphere abilities. Thus, it would appear that these individuals are severely intellectually imbalanced or lateralized. In each of these cases, it seems they are lateralized to the right.

This should not be too surprising. There are many studies that report a higher incidence of left-handedness among cerebral developmental disorders such as autism, dyslexia, and stuttering. Is it possible that autism, dyslexia, and savants are different aspects and degrees of the same aberrant development in cerebral lateralization, a relative hypofunction of left hemisphere when compared to the right? Could it be that they are all related to a common disorder of development?

Some may have difficulty with music skills being related to the right hemisphere of the brain. However, CAT scans of the brain taken on idiot musical savants often show focal lesions in the left hemisphere. A good example is described by Charness and Clifton (1988). They described John, a boy born premature and blind (retrolental fibroplasia) and severely mentally retarded. John had profound skills at the piano; however, because of a spastic right hand, he played with his left. An EEG showed seizure activity on the left and a CAT scan of the brain revealed left hemisphere damage.

As mentioned in chapter 1, Goldman reported in *Science* (1978), that surgical removal of parts of one hemisphere of the prenatal monkey produced a more advanced growth in the corresponding opposite hemisphere. It was found that this opposite side had a more extensive neuronal network, including connection with the opposite (damaged) side of the brain. Thus, if a part of one hemisphere is damaged, there is a tendency during development to compensate with increased development to the opposite side. It is as if the good Lord were to say, "Well, with that degree of damage to the verbal left side, he's not going to make much of a poet or playwright; therefore, let's give him some skills on the other (right) side so he can at least make a good engineer or architect." Is it possible that there is a law of relative conservation of skills between the hemispheres? What we subtract on this side, we add to the other, the total thus being the same—sort of different but equal.

Is it also possible that some of these skills are a form of subconscious logic at a more accessible level? We're not talking of ESP, but rather of an inborn skill. It is very interesting that savants don't know how they arrive at an answer. Somehow their subconscious "grinds" out the answer as a computer might. However, why are the calendar brain and the music brain (piano) so common among savants?

One of my nonneurologist friends once asked me if these amazing abilities were somehow related to ESP. Although some have reported a relatively increased incidence of ESP in idiot savants (Rimland, 1978), I think not. The reason is that ESP is an ability to receive data from distant areas. For example, a little girl is struck by a car. The mother, miles away, suddenly has a feeling of fear at the same time, calls home, and discovers the news. What we are dealing with in savants is a highly specialized skill in the processing of data. It is interesting to note that with the savant pianist, there is often early exposure to the piano. Often, there is a piano in the house where they live. They did not just walk up to a piano one day and play perfectly. It might only need a few minutes of exposure to play all the notes in order to get all the information needed. It would appear that when they first made noise (not music) on the instrument, their brain took in the physical location of the key on the piano and its correlation to the perceived audible note. Then more highly sophisticated calculations were made to develop chords, perhaps again by random cross-correlation or perhaps by a much more sophisticated means of Fourier analysis (mathematical algorithm) to develop chords from individual notes. All this was done within the confines of the subconscious, a sort of subconscious logic. Savants do not create music; rather, they reproduce it. It is a sophisticated processing of acquired data in a highly specialized manner.

What about the calendar brain? Again, first exposure to the data is needed to formulate results. As children, one of the most basic facts of life that we are all exposed to is the Gregorian calendar, which repeats itself every twenty-eight years. All the data needed to formulate is the concept of the seven-day week, the fifty-two-week year, the twelve-month year, the number of days to each month, the four-year leap year cycle, and today's date, including the day of week. This is all the information needed. The savant does not know how he formulates the result because the highly sophisticated process takes place in the subconscious. The savant merely looks up for a moment then gives the answer. There is no moment of deep thought, no emotion of uncertainty, just the simple and correct answer.

I believe these abilities function through the same principles of the experiences I had with subconscious logic as an engineer working with computers and as a sailor at night. Perhaps you have also had similar experiences. If it happens to a woman, we call it intuition; to a man, inspiration. It's all the same. Perhaps everyone has at least one brief period of genius in their lifetime. Is this a skill that we can nurture and build upon? Is it possible that we lose some of this skill with adulthood with its ever-present material demands, that we miss the forest for the sake of the trees? Is it possible to cultivate this skill? If so, should our present educational system take this into account? These are questions that are difficult to answer at present, but there is some interesting evidence to ponder.

The Cultivation of Subconscious Logic

Dr. Bernard Rimland reported in *Psychology Today* (August 1978) an interesting story regarding a normal graduate student. Ben Langdon. Two researchers wanted to see if a person could learn to calendar calculate and Langdon was selected to try. A one-page formula was developed and Langdon would attempt to do the calculations in his head. After trying day after day, he still could not match the skill or speed of the calendar savant, even though he became quite good. Then after many days, suddenly Langdon noted that the answer would just pop into his head. He became a calendar calculator and quite natural at it. It would appear that the calculations became automated in the right (nonverbal) hemisphere and were then delivered to the left (verbal) hemisphere for communication with the outside world. This concept of right hemisphere involvement is further enhanced by the aberrant communication between the two hemispheres sometimes noted on the few autopsies done on savants, such as the genius of Earlswood previously mentioned. Perhaps intuition and inspiration work in the same mechanism as the savant in that both use a form of subconscious logic.

You might ask if there is genius in you? The answer is yes. As we have discovered, the skills of the savant appear to be the result of an aberrancy of cerebral lateralization (to the right). This is the same lateralizing process involved in dyslexia, stuttering, and human behavior. If the savant is an extreme end product of this same mechanism, then there is a little genius in us all and it could be cultivated.

"Let Me Sleep on It Tonight."

As an electronic engineer, I lived in a beautiful place by the shore in southern California, Palos Verdes. Often I had difficult decisions to make or problems to solve. Palos Verdes has beautiful cliffs one hundred to one hundred and fifty feet above the shore of the Pacific Ocean. When I lived there, there was a quiet path that led down to the rocky shore below. Many times, when I had a difficult problem to solve, I would walk this path and then along the rocky shore toward San Pedro and the L.A. Harbor breakwater, just me, the waves, the dark clear night sky, and an occasional seagull or sea lion. I would feel at peace and the clouds of conscious turbulence would always lift; it was then that the solution to my problem would clearly present itself. I always went home with a solution or an answer.

I believe that our brains were designed to require regular periods of rest and relaxation in order to function clearly and efficiently. This is why God created the beautiful sky with clouds, the lofty green trees, the flowing waves of grass, the majestic mountains, and the powerful sea. We are at peace when under their influence; thus, the clouds of conscious turbulence are lifted.

One could carry this a step further. Have you ever noticed that after sexual orgasm solutions present themselves and difficult problems suddenly become clear? Sex is just as much a part of divine creation as the birds, trees, and seas. Perhaps we were designed to have regular sex as well as regular sleep. Even after a good night's sleep, solutions often present themselves to me. No doubt, this is where the phrase, "Let me sleep on it," came from.

Even in Scripture, God reinforced periodic rest by incorporating the seventh day of rest into the commandments. Who knows us better than our designer? You see, it was only when Langdon was able to relax that his brain was able to do calendar generation. The skill was there; all he needed to do was lift the clouds of conscious turbulence and the answer, generated in subconscious logic, was then able to penetrate conscious thought and present itself.

Thus, as educators and parents, we can encourage creative thought processes in our children by providing safe, peaceful, loving harbors that will lift the clouds of conscious turbulence, place the child's brain on a higher Maslovian level, and induce the creative process. Nothing kills creativity more than the fear of being wrong. Even management today is realizing this.

Subconscious Logic in Scientific History

Hooke. Evidence of subconscious logic is also present in the late seventeenth century, during the high point of the scientific revolution. As we all know, Isaac Newton discovered the law of universal gravitation, which he wrote about in his book *Principia*. Here he mathematically proved how all bodies in space are attracted to each other and thus orbit about each other. Newton proved that the force of these attractions varies inversely with the square of the distance. However, it is important to note that this formula did not originate with Newton.

Robert Hooke was secretary of the Royal Society of London in 1679. On November 24 of that year, he sent a letter to Newton in which he stated that he felt that the force drawing a planet toward the sun varied inversely with the square of the distance that separated them. Hooke's intuition, or inspiration, was later to be demonstrated correct by Newton on mathematical grounds. Hooke simply felt that he knew the answer, even though he did not have the mathematical genius of Newton to prove his point.

This is not to belittle Newton's discovery. Without Newton's genius, there was no proof, and without proof there was no real understanding. He was, perhaps, the greatest scientific genius of all time. However, by the same token, let us not forget the inspiration of Robert Hooke. His genius, although not expressed through the same vein and perhaps not to the same degree, was still genius. His genius (subconscious logic) caused him to be so impressed about the answer to the problem that he wrote a letter to Newton. We should admire Newton but not forget Hooke.

Edison. The story of Edison and how he invented the light bulb is another example of subconscious logic. Thomas Edison spent thousands of hours trying to find the right metallic substance to function as a filament in a vacuum glass tube. However, when he applied an electric current, the light always died out after a few seconds. Many so-called experts at the time started to feel that the concept of an electric light bulb was a red herring and not physically possible. But Edison persisted. He just knew it was possible. Then one day he wondered why use a metallic substance as a filament? He decided to coat a strip of thread with carbon and use that instead. Society has never been the same since the birth of the light bulb. Edison had no clear explanation for this line of thought; it was just an idea. Or was it? Could this also have been a form of subconscious logic at work in much the same way as Hooke, Langdon, or any of the *talents of the savants* showed? Perhaps inspiration is an expression of the hidden genius in us all and can be cultivated. One thinks of Luke Skywalker and the force in the movie *Star Wars*.

The Savants, as an Aberrancy of Lateralization

Is an idiot savant a separate entity or is he the result of the same pattern of defective neurological development noted in the dyslexic and autistic child? Is it possible that all these defects are related? To understand whether this is possible, we must learn something about these interesting disorders. Because of the higher incidence of savant skills in autism, no discussion of the savant would be complete without briefly mentioning this unusual developmental disorder. In fact, we now often hear the term *autistic savant* because it is gentler sounding than idiot savant and perhaps more accurate.

Autism

Autism was first described by Kanner in 1944. He described a withdrawal from social interaction and a lack of language development in these patients. Sometimes, he noted, the abnormality is seen early on in the first six months. Others might not show their abnormalities until the second year of life. Autistics appear mentally retarded at first contact; however, upon closer observation, their uniqueness is soon discovered. There are several characteristics that set autistics apart.

1. A lack of emotional development. These patients are withdrawn and keep to themselves. They do not develop social bonding with other individuals, as is often seen in mental

disorders such as Down's syndrome. They also have very poor eye contact.
2. There is relative preservation of motor development.
3. There is an unusually heightened sensitivity to the senses, such as smell, sight, and sense (feeling).
4. They have extremely poor communication skills. Sometimes they can read better than they can communicate verbally. However, their overall verbal skills are often almost absent.
5. They show signs of autostimulation. This includes handflapping, rocking, and head banging. They seem to be attracted to things rather than people. They are often preoccupied with spinning objects or other mechanical devices. They live in a world of their own. Often they will shake their hands in front of their eyes against a lighted background.

Several possible causes have been given to explain these characteristics, such as high serum serotonin levels or a defect in the reticular activation system (the brain stem system responsible for initiating and maintaining wakefulness and attention). Others have said the limbic system is implicated. None of these theories has been absolutely proved.

Autism and Cerebral Lateralization

Could a defect in cerebral lateralization be partially responsible for autistic behavior? If so, what is the evidence? First, like the idiot savant, there is a higher incidence of autism in males, which may further indicate a defect in left hemisphere development, possibly secondary to testosterone changes in utero. In fact, when reading speeches by Dr. Down from the turn of the century (Dr. Down coined the term *idiot savant*), he stated that there were *no* female savants. We now know that about five to ten percent of savants are female. Second, a preoccupation with things mechanical may indicate a left/right imbalance, favoring the right hemisphere. Third, a consistently profound defect in verbal skills suggests a relative dysfunction of the left hemisphere.

What about Emotion?

What about emotion? Now, one might think that a wrench is thrown into this mechanism because traditionally the right hemisphere is thought to be the emotional side. Then how could these autistic patients be less emotional in their social bonding if there is a relative increased function of their emotional right hemisphere? We shall see that the right hemisphere is not necessarily the emotional side. Emotion means different things to different people, which causes confusion in scientific study. I will explain.

What Is Emotion?

The concept of emotion and its possible involvement in cerebral lateralization is not well understood. First of all, the word *emotion* means different things to many people, including love, anger, explosive personalities, or hyperkinetic personalities. Stoic or emotionally solid persons might be called emotionally sound. It is generally known that the cerebral cortex has a regulatory (inhibitory) effect on emotions such as crying or inappropriate laughter. A good example of this is in multiinfarct dementia. In this disease, the cortical influence over deeper structures of the brain is being slowly eroded because of multiple small strokes in the white matter that connects the outer cortical gray matter with the central (emotional) gray matter. Thus, the central gray matter is free from cortical inhibition and the patient becomes very emotional. Neurologists call this a release sign. Often when examining patients, I use the following test. I will say, "Boy, did you see it rain the other day?" In severe cases, patients will burst into tears. Then I say, "But you know, the rain brings pretty little flowers." Then they quickly reverse themselves and start laughing. There is no real sorrow or happiness; it's simply a less modulated mechanism. It is under less cortical influence. Is this to be considered emotion? Or does one need to have the feeling behind the outward display for it to be called emotional? One might say it is "smart" to be in control of one's emotions. A veterinarian once told me that the best way to select a smart, trainable puppy is to pinch the paw. If its first reaction is to bite, don't buy it. However, if it licks first, that's the one to get. You see there is a difference between responding to stress and simply reacting to it. Responding is a much more cerebral and, therefore, more controllable mechanism.

Right Hemisphere as Modulator of Emotion

Which hemisphere is the emotional hemisphere? Many have felt that the right hemisphere is responsible for emotion by way of paraclinical tests and pathological findings. However, I have often felt that some of these might be a bit misinterpreted. For example, Lee and Loring, in *Neurology* (1988), report an interesting finding from a neurological test (Wada test) used for determining the lateralization of language and memory formation as a preparation for epilepsy surgery candidates. A hemisphere (left or right) is temporarily rendered nonfunctional by way of a locally injected (intracarotid) sodium amobarbital. There are two carotid arteries (left and right) in the neck. Each supplies the respective hemispheres with blood. Thus, if amobarbital is injected in the left side, the left hemisphere no longer functions (for a few minutes) and the person is only functioning with his/her right hemisphere. At that time neuropsychiatric testing (language and memory) is done to determine which hemispheres (temporal lobes) are most important if removal is to be considered. (For example, you would not want to remove the right temporal lobe if you were to discover this lobe in this patient to be important in new memory formation.) They reported that if a lesion is in the left frontal lobe, an injection of amobarbital to the right side of the brain would often result in behavioral disturbances. However, if a structural lesion were on the right frontal and the injection was done on the left, no such behavioral changes were observed. Thus, it would appear that there is a relationship between the right hemisphere and emotion. However, what kind of relationship? The authors interpreted the results as evidence of left hemisphere influence on emotion regulation (left hemisphere suppression of emotion). I do not agree with this interpretation, although the study was excellent. I believe that this study suggests the right hemisphere (not the left) to be the regulator of emotion. I will explain. . . .

They describe physical and verbal aggression resulting from right carotid injections: thus, it appears that the right hemisphere might have a regulatory effect on emotion. When it is temporarily put out of commission, there is a tendency for emotions to run free. Although there is a relationship between the right hemisphere and emotion, it would appear that it is one of suppression and not enhancement. For example, suppose that if, as suspected for many years, emotion regulation (suppression) originated from the cortex on both sides and this influence originated more from the right side than the left, then clearly, a known lesion on the left frontal side and temporary dysfunction of the entire right side would create more interference with the normal inhibitory function of emotion than injection on the left with lesion on the right frontal. It is a case of simple mathematics. There is "more bang for the buck," with the right-side injection. Therefore, it is the right hemisphere that has the greatest inhibitory influence on emotion, and if temporarily suppressed, there will be a greater emotional outburst. Could it be possible that the visual impression we had in college of the mechanical engineering major with a humorless dull expression, crew-cut hair, and slide rule in hand might be accurate? If you've been to college, you know of whom I am speaking. Could it be that his overfunctioning right hemisphere was suppressing his emotions? He was not trying to act that way, it was just his nature. Others have also reported that it might be the role of the right hemisphere to function as a graded and subtle modulatory agent on affective (emotional) expression (Ross, 1985). Therefore, my view of the right hemisphere, as a modulator of emotion, is supported by others.

Is autism the result of a gross extreme in lateralization of skills to the right (sacrificing the left)? Is the savant also an expression of that lopsidedness? Are autism, dyslexia, hyperactivity, and genius the results of a common lateralizing mechanism that takes place prior to birth? Perhaps. Much more investigation is needed to even begin to scratch the surface of what remains to be known of these conditions. It is an important field to investigate, for if true, there is an undiscovered genius locked in all of us waiting to emerge with the right key.

The Right Hemisphere Genius

Another excellent example of right hemisphere skills and their relationship to scientific genius is Carl Friedrich Gauss, who was one of the greatest mathematically gifted persons in history. He often joked that he could count before he could talk; thus, a subconscious hint to his right hemisphere skills. Gauss was born on April 30, 1777, in Brunswick, Germany, to a working-class family. His natural ability for mathematics soon became evident. One day while his father was calculating the weekly payroll for the workers under his charge, Carl was looking over his shoulder. Unaware of his son, Mr. Gauss suddenly heard his voice say, "The reckoning is wrong; the sum should be . . . " His father recalculated the sum and little Carl was proved correct. The interesting part to this story is the fact that Carl was only three years old and had not received instructions regarding arithmetic. Again, we have evidence of the processing of data in the right hemisphere in a highly specialized way, a sort of subconscious logic perhaps.

Another, more descriptive, example of his right hemisphere skills is this story, commonly told in advanced mathematics courses. Carl, a young boy of ten years of age, was in his arithmetic class. The class master presented a problem. What is the sum of $1 + 2 + 3 + 4 + 5 + 6 \ldots + 100$? The students were to figure the answer on their slate boards and then place the boards on the master's desk. They were to be stacked so that the first to figure out the correct answer was on the lower level (one on top of each other). Of course, the master knew the simple trick (formula) to solving the problem and thus did not need to add all the numbers. The master had just finished describing the problem, when suddenly Carl presented his slate to the desk with the single correct number on it. How was he able to do this? There were no calculators in 1787 and even if there were he could not have done it so quickly.

In his *Disquisitiones Arithmeticae,* published in 1801, Gauss

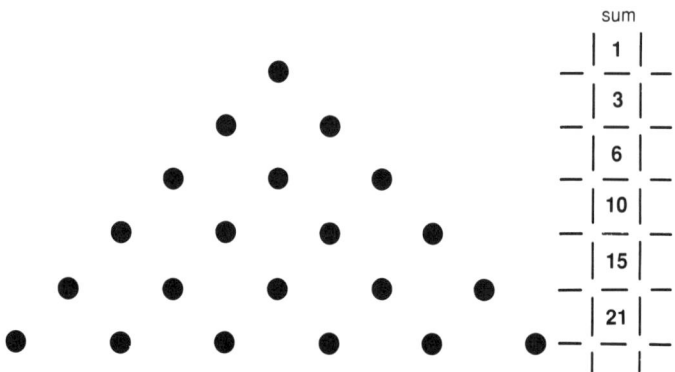

Fig. 18. Triangular numbers: N (N + 1)/2.

gives us an insight into his thought processes. In this work, he presents the concept of "triangular numbers" presented above. Each dot represents a single number 1 (integer). Gauss found that the sum of all the dots in a given triangle is the end result of 1 + 2 + 3 ... etc. for each level. The formula for solving the problem is thus $n(n+1)/2$. Therefore, the answer to the problem presented by the schoolmaster is $100(100 + 1)/2 = 5050$. Gauss was able to "see" with his mind and thus formulate the answer, which is typical of right hemisphere skills.

Discussion

There are the two kinds of highly developed skills in the right hemisphere. The analytic (proving an answer) and the inspirational (subconscious logic). Both may be an expression of genius. While most of us may not have the analytic type of genius, many may have experienced the inspirational type of genius. Perhaps genius is a focal islet of skill, while intelligence is more global in its cerebral functioning. Perhaps there is a little genius in us all, be it from left or right hemisphere.

No one hemisphere is the hemisphere of genius, logic, or creativity. Both sides contribute skills, be it the left (verbal) or the right (nonverbal). There is verbal and nonverbal creativity. Some authors describe the left hemisphere as the logical side. However, it might be because of our natural verbal/temporal world that one is tempted to make this relationship. However, in this chapter we have learned that logic can be verbal or nonverbal.

Is it possible that savants, autistics, and dyslexics represent different degrees of the same lateralizing process? We discussed evidence of a tendency of conservation between the two hemispheres, as described by Goldman (1978) and his experiments with prenatal monkey brains. We have also discussed the possibility that this skill may be cultivated. In chapter one, we discussed Orton's discovery (1924) that many of those with dyslexia were able to read much better looking through a mirror and how this might be related to a relatively higher functioning right hemisphere. In light of cerebral lateralization, we see the possibility that autism, dyslexia, and idiot savants may be the result of a similar process of aberrant cerebral lateralization that tends to slightly favor the nonverbal right hemisphere at the relative expense of the left.

6. Sex, Behavior, and Lateralization

This chapter brings together the many personalities and behavior patterns that are the result of cerebral lateralization. We learn that genius may be a focal area of skill while intelligence may be a more diffuse form of cerebral functioning. We see how nature favors diversity. A description of the right brain and left brain person is given.

I have always been a great observer of people, believing that by observing behavior we can gain insight into the inner workings of the brain. I consider myself something of a yachtsman. I'm at my best when sailing a large sloop in the Atlantic or in Chesapeake Bay. Every fall, when able, I attend the sail boat show in Annapolis, Maryland. A most interesting human behavior takes place there. Couples sometimes do a very curious thing when coming aboard. The female often goes below first and views the woodwork and the layout of the interior. The male will often first check out the rigging topside and comment about that. Is this a clue to understanding ourselves? Some may argue that this behavior is somehow the result of social conditioning. This might be true in some cases. However, what is apparent is that this behavior often seems to be their first impulse, just after boarding the boat. Is this behavior a clue to the hardwiring sexual differences of the brain? All sailors will acknowledge that no yacht is complete without an adequate interior and exterior. This way the yacht becomes whole. Is it possible that through this behavior we see another indication of the two halves of our brains becoming whole? One half of the couple covers one aspect and the other another and by this we have a complete whole team able to do great things. No superior or inferior, just two halves becoming whole. Every time this is observed, I get a feeling of completeness and wonderment about this great design of nature. There is much more than just physical attraction that draws the sexes together.

Some of my biologist friends have argued that this functional attraction is the result of a biological need to maintain a state of wholeness seen during the early evolutionary stage of asexualism hundreds of millions of years ago. As if to say that when the sexes were formed, each sex took with it a unique entity, and that this attraction is the desire of the particular sex to get that back. Deep stuff to say the least. However, there appears to be purpose, logic, and utility to the formation of the couple. This is not a question of superiority or inferiority but of wholeness. Psychologists have known for a long time that even among homosexuals there is often a profound difference in behavior between the partners. Each has different duties and responsibilities, allowing the couple to function harmoniously. Thus again we have a sort of difference with equality.

The Brain's "State of Mind" and Lateralization

Lateralization in the Relaxed Brain

Is it possible that various skills in the brain help determine some of our behavior patterns? The concept of cerebral lateralization teaches a mechanistic approach to the brain; that much of what we are is the result of hardwiring that began prior to birth. This hardwiring can result in providing various skills unique to either the left or right hemisphere. It would appear logical that if someone were blessed with a given brain skill, there would be less stress in exercising that skill and thus an influence on behavior. Here is a theorem that I have developed, which explains the driving force of behavior in the lateralized brain.

Cerebral activity in the relaxed lateralized brain will tend to shift (lateralize) to the side (left or right) that has been endowed with the greatest skill.

The key word here is relaxed. It takes less effort or stress for a brain to perform a task in a gifted area as opposed to a less gifted area. By desiring to relax, we naturally want less stress and therefore shift to the more developed easily understood subjects. Thus, another example of behavior following skill. Let me give an example....

A couple relaxes during a pleasant afternoon. Both want to relieve stress. The wife goes upstairs to the bedroom to watch "Days of Our Lives" on TV or read a novel, while the husband spends the afternoon in the dining room building a model of a ship. Is it just chance that each sex relaxes this way or is there an underlying basic neurological mechanism at play? Is it just by chance that the female desires to relax in the left hemisphere (verbal/temporal) while the male relaxes in the right (visual/spatial) side? Often when I question women why they view soap operas, they say, "Because it makes me relax." I have always thought this was interesting because after trying many times to get involved in these stories, I just can't relax. The stories are often fairly confusing and require a good deal of concentration on my part. I feel much more relaxed using my right hemisphere building a model ship.

Even in the case of idiot savants with amazing musical skills, they truly love to perform, even if left alone. Perhaps their musical product is the result of a burning desire, which is again the result of a lateralized skill determined at birth. Behavior following skill.

Lateralization in the Stressed Brain

One could conclude that if behavior follows skills provided by a particular hemisphere, then there is a tendency for that person (left hemispheric or right hemispheric) to view his/her world through that particular hemisphere.

The lateralized brain will tend to perceive the world as well as solve problems through the same lateralized hemisphere when under stress.

We saw evidence for this with the story of the maps and directions in the second chapter. The left hemisphere person will tend to verbalize direction while the right will draw a map. Both work just as well for each perceives his/her world through their lateralized hemisphere.

Now, okay. This seems logical. But with all the marvels of modern scientific technology, is there any technological evidence of this process?

There was a very interesting study done in West Germany by Landwehrmeyer and Gerling (1990) using a special EEG that shows cortical negativity. It is felt that when a part of the brain cortex is used to solve a task, there is an increase in negativity there. Landwehrmeyer and Gerling used ten normal subjects and ten with dyslexia. They found that with the controls (normal subjects) there was an increased negativity in the left hemisphere during reading. This is not surprising because the left hemisphere is considered the side for reading. However, with the dyslexic subjects, there was a tendency for increased negativity to the right hemisphere during reading. This suggests that the lateralized brain (in this case having more skills on the right), when stressed, will tend to solve problems with the hemisphere that has the greater skills, even though the other hemisphere is specialized for that task.

This finding would be in agreement with Orton's report (1925) that children with dyslexia were able to read better when viewing through a mirror. The mirror caused them to read right to the left and therefore the right hemisphere, having higher skills, was used to lead on the next letters and words.

Thus the left hemisphere person lives in a left hemisphere world and a right in a right world. This lateralization of skills affects not only how we perceive our world but how we try to handle the various problems our world presents. This should teach us tolerance and understanding with others that seem different, for we are all different but equal.

No Room for Prejudice

Cerebral lateralization is not to be used for prejudicial conclusions. Any female may be just as right hemispheric in thinking as any male and any male as left hemispheric as any female, for there are vertical differences as well as horizontal. We should not jump to conclusions about a particular person based on data from large groups. The logic does not always flow in both directions. Remember, *All squares are parallelograms, but not all parallelograms are squares.* Judy should be able to be anything she wants and the same for John. However, we can use this new knowledge to help understand each other.

As an engineer, I remember observing how globally intelligent female engineers were. Not only were they good engineers, but almost all were good communicators. Their command of the language and spelling was usually perfect, often in great contrast to their male colleagues. It was as if nature endowed them with right hemisphere skills in addition to their gender-advantaged left hemisphere skills; they were thus globally intelligent.

I recall visiting two male patients living in the same room at the hospital. They had been convalescing for several weeks, and both men were always busy building model cars, ships, and planes. When asked why, one replied, "It beats watching 'As my Stomach Turns.'" I was not sure what he was alluding to, but I felt fairly certain it was a soap opera. Often men will discredit women who enjoy viewing these shows, but perhaps it's a defense mechanism because they live in a right hemisphere world and do not feel comfortable trying to relax in a world where their skills are less developed.

Why Sexual Attraction?

Even though the sexes are quite different, they are still attracted to each other. Why is this so? One answer might be that it is the difference that causes the attraction: "Viva la difference." Is it possible that natural desire for the union of the sexes is the result of functionalism and is not just merely physical? Could marriage and romance be the merger of two minds, a mutual symbiotic union of two spheres of skills, resulting in wholeness, completeness, and security? There is something that you can't explain about the other person that we find missing in ourselves. It is my hope that this book will be helpful in explaining that something. We should not try to change our partner's behavior but rather enjoy and appreciate the differences between us. It's not a question of superior or inferior but of difference.

A simple illustration of the difference between the two sexes in processing data is the physics of television. I have believed for a long time that by understanding all aspects of science, we understand a little more of ourselves. This includes the hard sciences, such as mathematics and physics, in addition to life sciences and social sciences. You may remember that one almost *expects* a woman to remember patterns and colors of a room, while a man may not. The woman might say, "The walls were an awful pink that just didn't match those tacky drapes." On the other hand, a man might recall an interesting shape to the legs of a chair. Could physics help us understand this? When I was in college, I used to make a little extra money repairing televisions. There is a principle in basic television physics that I never forgot. You see, in order to transmit a television signal, you need a spectrum (range of frequencies) from one to four megahertz wide. (One megahertz equals 1 million cycles per second.) Now something interesting happens to a picture if only the first half (1–2 megahertz) of the spectrum is used; the only thing you will see is nonspecific forms and colors. Much of the shape of the object is lost, but there is good color definition. Now if only the higher half of the spectrum is used (3–4 megahertz), the colors will be lost but the forms (outlines) will be well preserved; the shape of things will be well defined. Perhaps there is a key to understanding ourselves. One may think of the first half of the spectrum (1–2 megahertz) as a more female view in which colors are rather important. The latter half (3–4 megahertz) might be considered to be more of a male perspective, where the shapes become impressive. When looked at separately (higher or lower spectrum), the picture is lacking significant visual information. When the whole spectrum is used, the picture becomes whole and complete.

Perhaps, this might explain why marriage is an advantage to completeness or wholeness of life. Your mate sees a different world that has just passed you by. Your first impression might not be the same as his or hers, but by being together the picture is complete and little is lost. There is wholeness and completeness here, a coming together of the hemispheres to make a complete entity. It just might be possible that through the union of the sexes there is a wholeness that allows for a complete perspective on life, a thoroughness that otherwise is lacking. Perhaps that "certain something" in the mate that attracts us may be more than just physical; it might be that other hemisphere, that other half that we find lacking. After all, you don't need a Ph.D. in physics to know that opposites attract.

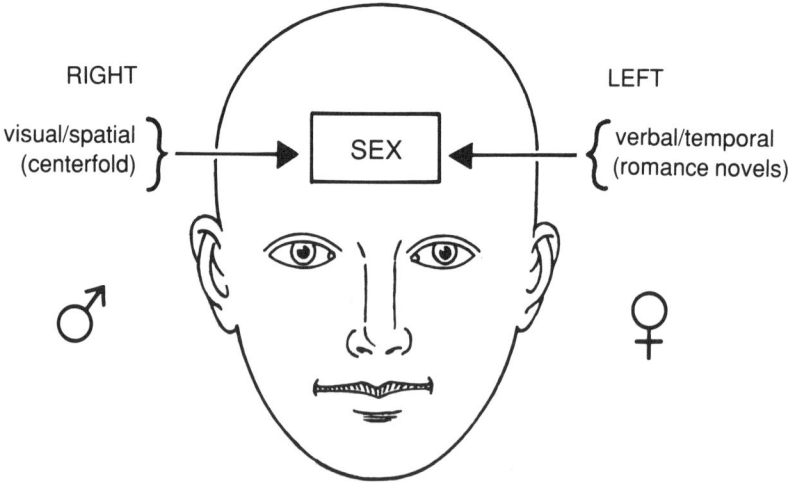

Fig. 19. Male versus female sexual stimuli.

Let me explain in another way how the sexes can be different and yet alike. Sex is of mutual interest and enjoyment to man and woman alike. Yet to achieve sexual gratification, each sex has its own preference for stimulation. It is interesting to note that we

112

see this difference of preference everyday at the drugstore and the supermarket. Next time you go, look at the sex-related magazines. For the male we have *Penthouse, Playboy,* and *Hustler,* all of which deal with sexual stimulation through the right hemisphere (visual/spatial). For the female we have *Cosmopolitan* and various romantic magazines and books, all of which deal with sexual stimulation through the left (verbal/temporal). Some people in the publishing business have known about this difference for years. Not too long ago, *Cosmopolitan* tried a centerfold for women showing nude men. However, this never became very popular and was stopped. I feel that there will always be some centerfolds for women, but I doubt they will ever be as popular as they are with men. Thus, it would appear that men and women may have different paths to the same object of sexual gratification. Needless to say, there are different degrees of lateralization between the sexes. Not all women are 100 percent left and males 100 percent right. Perhaps many are 60/40, some 70/30, and a few are 90/10. This is a personal preference. However, there appear to be generalities that we can learn and benefit from. Sex therapists have known for many years that among women's most common complaints about their male sexual partners were: (1) lack of sensitivity (verbal stimulation) and (2) lack of foreplay (sexual stimulation). These things are obviously very important to the sexual gratification of the female. Males might complain that the women rarely wears anything sexy. Thus he may feel that his path to sexual gratification is partially blocked for lack of visual/spatial (right hemisphere) stimulation. Is there any biological data to explain or support this? Yes, there is.

Is Lateralization Skewed?

In chapter 2 I described how Diamond, Dowling, and Johnson (1981) reported a statistical difference in thickness in the hemispheres between adult male and female rat brains. Males' brains were thicker on the right, females on the left, which is perhaps not too surprising at this point. However, there was also evidence of a statistical difference in cortical thickness between the anterior (front) and posterior (back) areas of the rat brain. Males were thicker in the anterior (motor), and females were thicker in the posterior (sensory). So perhaps sexual lateralizing differences are skewed and are not just left and right. Perhaps by being skewed, a high degree of variation in personalities is possible.

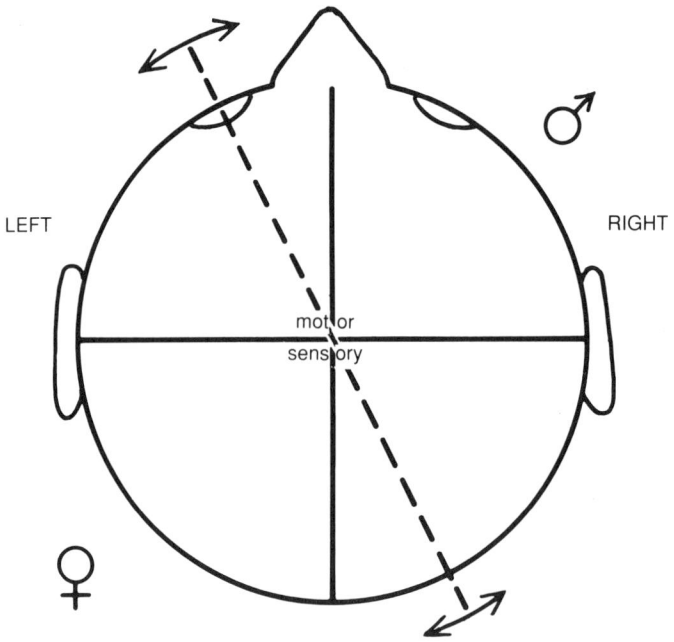

Fig. 20. Professions and lateralization: The bilateral (superbrain) and unilateral person.

As mentioned previously, no single gender has exclusive rights to a hemisphere. A female can be just as right hemispheric in lateralized skills as any male. Also, any male may be just as left hemispheric as any female. However, because of the apparent trends of sexual hormonal influence in lateralization, there are some interesting conclusions. For example, although there are fewer female electronic engineers than male, being an engineer I have had the opportunity of meeting several females in this profession. I have noticed that they often appear alert and well balanced in their skills. In other words, besides being gifted in math and science, they also generally have excellent communication skills. They could spell perfectly. You might say they were globally gifted with high-functioning skills in both hemispheres.

This is in contrast to their male counterparts who, although gifted in math and other visual/spatial skills, were often poor in communication skills (myself included). This is easy to explain when considering the concept of cerebral lateralization and its relationship to testosterone. By virtue of the fact that the females have higher left hemisphere skills, any added right hemisphere skill is just "icing on the cake." I often call this global hyperfunctioning a "superbrain." This is because the whole brain (both left and right hemisphere) is highly skilled. These people are very valuable in society, for they not only provide right hemisphere skills, but they can ensure a proper communication of ideas through their high-functioning left hemisphere. They may be involved in almost any profession; however, of particular interest would be math and science teachers, accountants, executive secretaries, computer programmers, physics professionals (theoretical), etc. Although any profession would benefit from a superbrain, these seem to benefit particularly from the diffuse skills of the globally gifted as they require good right hemisphere skills but good left hemisphere skills to communicate their ideas. I wish I could count myself among these globally gifted people, but because my right hemisphere skills are greater than my left (typical of those with a history of dyslexia), I cannot. I should also point out that I'm very left eye dominant, even though the visual acuity of both eyes is the same. Could this also indicate a cerebral imbalance favoring the right hemisphere? No one knows and much study needs to be done. Try the ocular dominance test yourself. So far with our study of cerebral lateralization, one might postulate that a female with right hemisphere skills and a male with left hemisphere skills could be a superbrain. This is not to be confused with genius, which may be more focal in skill, as described in chapter 5.

Other professions not as demanding of both hemispheres might be engineers, music composers, physicists, mechanics, for the right hemisphere; playwrights, novelists, general secretaries, speech writers, journalists, lawyers, etc., for the left. One might refer to these people as unilateral (one sided) in skill as opposed to the superbrain, who might be called bilateral (both sided) in skill. Of course, we are speaking of generalities, and each profession has its own unique demands, which may require skills from both sides of the brain. Neither is this to say that one is superior or inferior by virtue of profession but that different professions require different skills. Therefore, this might explain why it is difficult to find good physics teachers for our children. One can be bright, and have right-sided skills (typical of a physicist); but without adequate left-sided skills for communication, one will be worthless as an instructor. Thus, ideally, a teacher (professor) will be bilateral in skills.

What other evidence is there for unilateral types that we might see from day to day? Next time you examine the Yellow Pages of the phonebook, look carefully at the type of directions given in the various advertisements. A lawyer's office may very well have the directions written at the bottom in longhand (verbal), while an auto mechanic may have a little map (spatial). Each looks at their world through their own cerebral orientation (lateralization) and feels everyone else must view the world the same way, the mechanic with the right hemisphere and the lawyer with the left. But we are all different and are impressed by different things, and good marketing should take this into consideration. One could easily see that sometimes a combination of a map and verbal directions might cover a wider market, making it easier for all types of people to locate the business, thus increasing profits.

Even in my profession, medicine, there is a rather consistent behavior among the various specialties that deal with lateralized skills in the brain. For example, surgery is a typically right hemisphere field by virtue of its demanding visual/spatial concepts in anatomy. Now ask a surgical resident to write a history and physical on the hospital chart of a new patient and you will be lucky to get a single page. They are not interested in the verbal/temporal aspects but live in a visual/spatial world. Thus they write very little. Now let's look at the left hemisphere. Who could be more left hemisphere than a psychiatrist? Psychiatrists always deal with their patients' pasts (verbal/temporal) trying to find clues to the present. If you ask a psychiatric resident to do a history and a physical on the same patient, you will most likely get more than four pages. He lives in a verbal/temporal world, and this is what's important to him. Thus, he writes a lot and in great detail. It has also been my experience that women medical students almost invariably write a longer and more detailed history and physical than their male counterparts. Perhaps this, too, is a key to understanding cerebral lateralization between the sexes.

Behavior and Lateralization (Hyperactivity)

The idea that lateralization of brain function may have an effect on behavior is not new. David Baer and Paul Fedio reported (1977) on the different interictal (between seizures) behavior of patients with left versus right temporal lobe epilepsy. A total of forty-eight subjects were tested in four groups. Patients were rated on things such as depression, paranoia, religiosity, sexual alteration, emotionality, aggression, etc. They found that right temporal epileptics displayed emotional tendencies while the left temporal epileptics showed ideational (hyperreligiosity, etc.) traits. It was felt that this difference of behavior could be secondary to an imbalance between the hemispheres brought on by a lesion with secondary hypofunctioning of the same side as the seizure. This should not be too surprising to the neurologist because psychiatric disorders have for a long time been associated with many temporal lobe epileptics. This has been noted as far back as 1963, when Slater and Beard reported on schizophreniform behavior and epilepsy. Could it be possible that some psychoses are the result of an extreme imbalance between the hemispheres, with the other end of the spectrum being simply a mild difference in behavior? No one knows as yet.

Is there any clinical evidence for the lateralization theory of behavior? There are some fairly predictable clinical syndromes that may suggest so. I will give you two settings that illustrate this from opposite perspectives.

Interictal Behavior and Anticonvulsants

Neurologists have known for a long time that these (interictal) behavior changes often become more frequent when the seizures are under good control with anticonvulsant therapy. What could the seizure medicine be doing that would allow it to tend to bring on psychotic behavior? Taking into consideration that the temporal lobe lesion that produced the seizure is causing a functional imbalance between the hemispheres, which might account for the intermittent psychosis, was suggested by the Baer/Fedio study; the anticonvulsants might be further enhancing that imbalance because of their cortical suppressant effect and thus aggravating the psychosis. How could this be? Well, actually it's mathematical. We know that one side effect of antiseizure medicine is a slight slowing of the cerebral cortex, thus, as the activity of both hemispheres is reduced, any previous imbalance will be more lopsided as a percentage. This is the law of ratios (see figure 21A).

We see that with the anticonvulsant medication, any imbalance is exaggerated; therefore, there is an increase in behavioral changes. One should not be surprised of an increase in psychotic-like behavior when seizures are under good control if one considers basic math and the lateralization of behavior theory.

Hyperactivity and Therapy

The syndrome we now call hyperactivity in children became relatively well known in the 1950s. At that time it was known as simply minimum brain disfunction, hyperkinetic syndrome, or a learning disability. Today we simply call it hyperactivity. There are some interesting facts about this disorder.

1. It is very common. Probably 5 to 10 percent of schoolchildren manifest symptoms of overactivity and inattention. This obviously interferes with their learning in class.
2. It is much more common in boys. The ratio of boy to girl incidence may be as high as six to one. Could this be an indication of the possibility of cerebral lateralization and its sex hormonal influence on this disorder? I will explain . . .

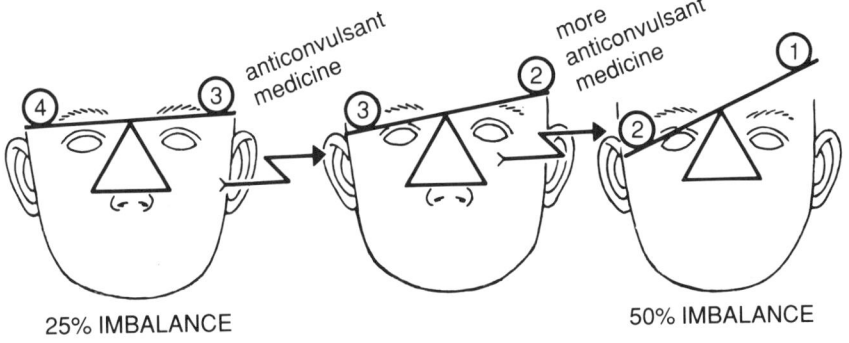

Fig. 21A. Law of Ratios (minus one each): Shows how cortical suppressants create lateral imbalance.

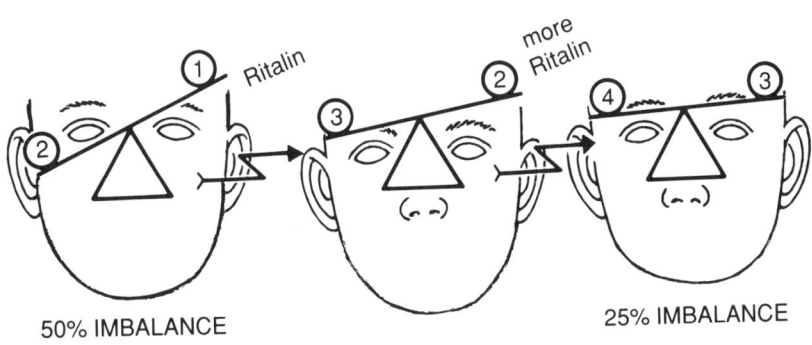

Fig. 21B. Law of Ratios (plus one each): Shows how cortical stimulants create more balance.

The hyperactivity in these children has often been felt to be secondary to stress avoidance. When confronted with an intellectual task for which there is a relative deficit of skill, the child becomes uncomfortable and stressed. The child will thus run away to avoid the stress and therefore will become hyperkinetic and inattentive. It is for this reason that hyperactivity and attention deficit disorder are sometimes used together to describe the same syndrome of hyperkinetic children, although they may be two separate disorders.

However, could this not be considered another example of behavior following skill, in this case, running away (stress avoidance) from a lack of skill? Thus the energy is not directed, which might explain the often destructive behavior of these children. I recall one of my hyperactive boy patients making a complete mess of my office. He threw everything on my desk on the floor. Considering that this is more common in boys, could there not be a sexual developmental component to the cause? Looking back at the sexual differences discussed in chapter 2, one could construe that it may be the result of a relative deficit of skill in the left hemisphere typical of the young male brain. (We will discuss later how there is evidence that the developing left hemisphere lags behind the right and that the male developing brain lags behind the female.) This idea is made further convincing by the fact that the demands on the left hemisphere are so great in the early years of life, both in and out of school. These demands include reading, writing, and arithmetic, in addition to typical left hemisphere verbal demands. Thus, the girls often do rather well during these years while the boys need time to catch up.

What further evidence is there that hyperactivity may be the result of asymmetric cerebral lateralization in the developing brain? Again we look at therapy, as we did with the temporal lobe epilepsy studies described previously. One could postulate that if hyperactivity is the result of a functional cerebral imbalance, then going in the reverse direction (taken by the previous epileptic study) and increasing overall cortical activity (cortical stimulants), the imbalance should be less and hyperactivity lessened.

As we see, the law of ratios again helps us understand

ourselves (see figure 21B). It is a well-established fact that cerebral cortical stimulants can be very helpful in treating hyperactive children. In fact, they are the most helpful of all drugs. These include several drugs such as methylphenidate (Ritalin), pemoline (Cylert), and dextroamphetamine (Dexedrine). Even caffeine can be of significant help in the management of these patients. Not all children require medication and this is a personal matter between the physician, the patient, and the family. However, there is little doubt that these drugs do indeed calm the patients down, even though they are also generally used to make people more alert.

Some investigators try to explain this therapeutic result as some unknown effect the drug may have on the motor systems; perhaps the motor system may have a receptor sight for the drug that is present only during the developmental stages. Others have implicated the reticular-activation system in the brain stem as the cause for their effectiveness. Nothing has of yet been proven. Researchers have tried but they have not yet found these drugs to have a biochemical effect on brain motor systems, and I doubt they ever will. However, there is one common denominator to all of them; they all are cerebral cortical stimulants. Could it be possible that they work by smoothing out the imbalance between the hemispheres, either directly on the cerebral cortex or by stimulating the brain through the reticular-activation system in the brain stem? Again we see clinical evidence of behavior following skill.

Summary

Lateralization and Education

Thus we see clinical evidence of behavior being affected by the balance of the cerebrum (lateralization), both in therapy as well as in psychological studies. We have also seen how this lateralization may be present in the sexes and how this difference may draw us together. The study of temporal lobe epileptics has also revealed some interesting conclusions about differences in development rates between the hemispheres as well as the sexes. Taylor (1969) reported, based on his study of children with febrile seizures who later developed temporal lobe epilepsy, that there is a difference of growth rate between the hemispheres and the sexes. He concluded that: (1) the left hemisphere matures after the right hemisphere, and (2) the male brain matures after the female brain. Therefore, it is not too difficult to see that if the male brain matures slower and its left hemisphere matures slower than the right, the young male may have some behavior problems that would not be seen as frequently in young females because of the slightly more pronounced imbalance between the hemispheres in the male. Therefore, there is a higher incidence of hyperactivity and similar disorders in males. They would also have more difficulty with the typically left hemisphere skills demanded in the early years of school. Patience is so important when dealing with children's education.

Educators today have learned to be slow to label anyone inferior, for when doing so, you may rob both the child and society. Today we know that education is not just the mere conveyance of information and the evaluation of performance. The greatest gift that an educator can bestow on a student is the thirst for knowledge. That way true education is inspired throughout life. For if there is no thirst for knowledge, much of the information gained will be lost and "all is for naught." Cerebral lateralization reinforces that concept. Education should be an opportunity for all, for the mind is far too complicated to be measured in a simple test and then deprived of learning because of a low score achieved during an unfavorable developmental period. How does one measure drive, inspiration, and creativity anyway? Perhaps in the balance of time, we are all different but equal. All adults should consider higher education regardless of the grades they made in their youth. Some other cultures might benefit from this idea, for I too was labeled inferior during a less enlightened time, simply because of dyslexia and hyperactivity and did not do so badly when given the opportunity.

7. Memory

Cerebral lateralization is a mechanistic view of brain function. Is memory, which is so vital to life, also compartmentalized in the brain as well? This chapter discusses how there are specific "data entry zones" for the formation of new memory and how recall can be triggered from specific sites on the brain.

Is it possible that man, being a religious animal, may store religious information differently? Does this have any spiritual implications?

Memory

When I was a freshman in engineering school, we had an introductory course in computer programming. The year was 1969 and computers were not commonplace as they are today. They were also rather primitive. However, we all knew that computers were the wave of the future and that an engineering education without them was less than adequate. Therefore, we were eager to learn. Our department did not have a large computer, so we shared the use of a Univac 1108 many miles away via a phone terminal and an electromechanical teletype. The set up was less than ideal but adequate to learn the language of scientific computer programming (Fortran).

Memory versus Learning

It was in this class that I really began to understand the difference between memory and learning. You see, as man first started to develop the concept of computer programming with advancing technology, he came to grips with the actual application of learning theory because memory is the mere storage of information while learning is the self-acquired application of that information to achieving a goal. I was fascinated to discover that computers can actually learn and that learning theory was an entire field in the study of computer software. I will explain. . . .

In the early development of computers, a game program was often played called pick-up sticks. I remember as a freshman playing this game on our noisy electromechanical teletype machines in college. The rules were that you were given an allotted number of sticks and were to play against the computer. Each of you was only allowed to pick up one or two sticks at a time. The one left with one stick was the loser. The computer was not aware of the rules but was told when it had won or lost. At first, it was very easy to beat the computer for it had no knowledge of the game. However, after repeated games, we soon discovered that the machine was no longer ignorant of the rules. Through a learning program the computer was able to learn the game and became impossible to beat. I was always stuck with the final stick. I found myself almost jealous of the machine's learning capacity.

It was at that time that I thought, perhaps a relationship or an emotional bonding between man and machine might some day be a problem if man lacked understanding of computer mechanisms, a sort of computer dating with androids. We've all seen this on the old "Star Trek" series. However, with our continuing advances in increasing memory storage capacity and computer speed coupled with the fact that these machines can learn, what will be the end result? Could you imagine the result of two sophisticated computers playing chess against each other, each with an equally advanced learning program? Considering their speed of operation and memory, the machines could be absolute masters of the game in no time.

Then another question comes to mind. If these machines are not only able to talk to each other but can learn from each other, if they can recall memory and learn as well as we do, how do we differ from them? Is the difference simply a degree of complication and sophistication, or is it more? I have often believed, even as a child, that to truly understand ourselves and who we are, we must attempt to become familiar with the brain's microuniverse and its mechanism. How do we process and store information?

The Processing of Data

It has long been known that memory appears to be stored diffusely throughout the cerebral cortical mantle (outer shell). There does not appear to be any one location where all memory is stored and recalled. I will discuss later how the temporal lobes seem to be intimately involved in recall of memory. However, there does appear to be a small area of the brain that deals with new memory formation. A sort of "bus line" for on-coming new data (verbal or visual) about to be stored in memory. All data to be stored in memory must pass through this data entry zone. The index of suspicion for the existence for these data entry zones has been high for many years because of a well-known clinical syndrome from alcohol abuse. The syndrome is called Korsakoff dementia and is the result of thiamine deficiency. Their inability to form memory is so profound that often they are unaware of their problem. They seem to forget that they forget.

For example, you may walk into the room to visit with such a patient and have a pleasant conversation. He may truthfully tell you his rank in the Korean War, as well as some other true stories. For all this is old, already acquired data. However, if you leave the room for five to ten minutes and then return, he would deny having ever seen you before or talking to you. This is characteristic of a total inability to store new information. Now what is interesting is that the brain lesions found on autopsy are very small and localized to the deep central structures, such as the thalamus and brain stem. There appears to be a data entry zone for memory storage. This reinforces the mechanistic approach to cerebral functioning found in cerebral lateralization. Where is this zone? Our brain has specific areas of cerebral function for data entry, and the fact that this area might be on a particular side of the brain (the left hemisphere) reinforces this concept. I will try to explain this theory by using a clinical history of a patient from my files who developed a syndrome called transient global amnesia. First let's describe this fascinating syndrome.

Transient Global Amnesia and Memory Storage

Transient global amnesia (TGA) is a temporary memory disorder that occurs not infrequently in middle-aged and elderly persons. The memory defect mainly involves the formation of new memory and the recent past. There is no impairment of consciousness and personal orientation as well, as other basic neurofunctions remain grossly intact. There are no clear-cut signs of seizure activity. Full clinical recovery usually takes place in twenty-four hours.

In 1986, Caplan suggested four minimally acceptable criteria for the classification of attacks as TGA. They are as follows:

1. The onset of attack should have been witnessed.
2. Dysfunction during the attack should have been limited to repetitive queries and amnesia.
3. There should have been no other major neurological signs or symptoms.
4. The memory loss should have been transient, from hours to one day.

What I found was a person with this syndrome (fitting Caplan's criteria) and having a left thalamic infarction (stroke) on an MRI scan of the brain. This supports the concept of a data entry zone for memory formation in the left thalamus. I am not alone in this concept as Gorelick (1988) previously published a case of a patient with this syndrome who later developed an infarction (destruction) of the left thalamus.

Description of the Patient

The patient was a sixty-five-year-old gentleman who presented to Georgetown University Hospital with a complaint of not being able to remember the events of Sunday mass on the day of admission. He stated that he remembered going to church in the morning but could not actually remember being there. He was brought to the emergency room of Georgetown University Hospital via ambulance because of the concerns of his friends. However, once in the ward, he could not remember being brought by the ambulance or the events that had taken place in the emergency room. There was no loss of consciousness, obvious focal neurological deficits, or any signs or symptoms of seizure activity or other obvious clinical changes. The patient was otherwise healthy and denied having a history of hypertension, diabetes, previous strokes, head trauma, or seizure disorder. He also denied having had any similar previous episode.

The patient was examined in the emergency room by a neurology resident. The patient was alert and oriented to his surroundings. He thought the month was April when actually it was May. His complete neurological exam was normal, except for memory formation. He remembered three out of three words immediately but was unable to recall them after three minutes. In fact, he could not recall being asked to remember the three objects. Every aspect of his physical exam was normal except for new memory formation.

Thus we have a classical presentation of transient global amnesia. As you can see, this patient fits Caplan's criteria.

An EEG was within normal limits as was a CAT scan of the head taken on the day of admission. However, the MRI taken on the following day revealed a small focal signal abnormality in the posterior portion of the left thalamus, measuring approximately two to three millimeters in diameter.

This lesion was seen on both the transactional and the coronal cuts (different angles of picture taking) and did not produce mass effect (was not a tumor). This is a typical finding in a lesion that is suspected of being a small stroke (disruption of blood supply).

The patient's condition improved completely and he was discharged after only a few days of hospital stay without complaints or neurological problems. This again fits Caplin's criteria for transient global amnesia. No medication was prescribed except for stroke prevention.

The Cause of Transient Global Amnesia

The etiology (cause) of TGA has been a mystery since its initial description by Fisher and Adams in 1958. At that time the authors speculated that the cause might originate with seizure. More recently mesiotemporal and thalamic infarction has been suggested (Kushner, 1985) as well as vasospastic changes, possibly related to migraine, as described by Hinge (1986).

Scientific History for Thalamus and Memory Storage

The concept of a thalamic lesion and pure memory formation deficit is not new. Victor and Adams, in 1971, postulated that the dorsal medial nucleus of the thalamus and the more posterior nucleus, including the pulvinar, may be crucial to the formation of new memory, based on their study of autopsies done on patients with Wernicke's/Korsakoff disease (alcohol-related dementia). Later in 1979, Squire described a well-known and remarkable case of an individual who had sustained a stab wound to the brain with a miniature fencing foil in 1960. The foil had entered Herman's left eye and punctured his left brain. As a result of this wound, he had been suffering from severe verbal memory deficits and had been unable to form new memory. In other words, if you were to visit him, leave the room for a few minutes, and then return, he would have no recollection of ever seeing you or the previous conversation. Many were at a loss for an explanation and imaging technology was very limited in 1960. Many years later, in 1979, the CAT scan became available and one was done of the head in this same patient. It revealed a small lesion, focal in nature, limited to the region of the left dorsal medial nucleus of the thalamus. Brain scientists had been speculating for years that this small area in the central part of the brain might be involved with new memory formation, and this report presented some hard evidence.

Heilman, many years ago, postulated that this same dorsal medial thalamic nucleus may be specifically important in the formation of new memory because of its reciprocal projections

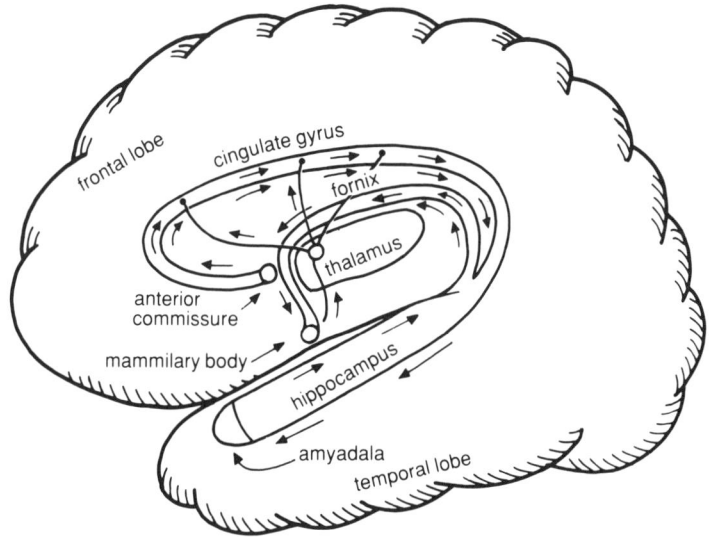

Fig. 22. Circuit of Papez limbus.

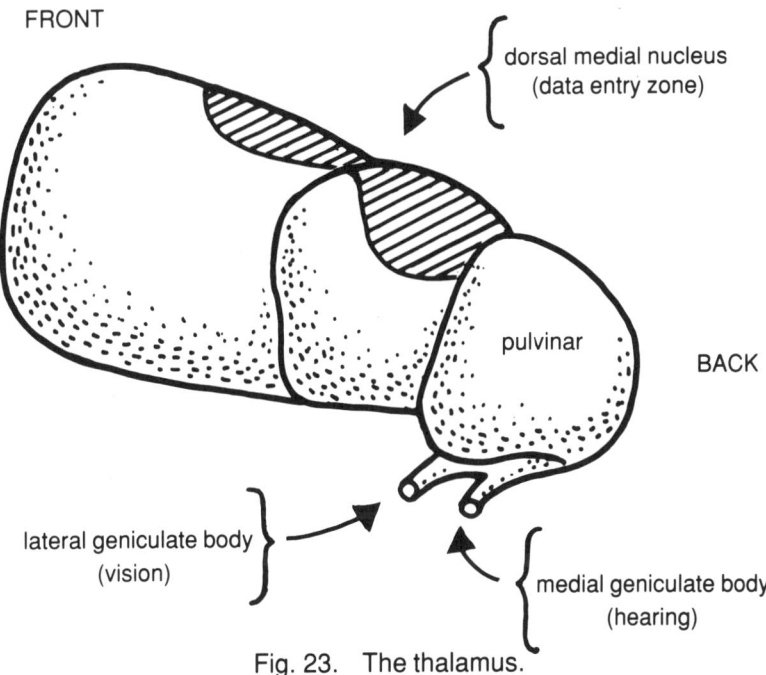

Fig. 23. The thalamus.

with the amygdala and thus part of two well-known circuits of the brain. In other words, this section of the thalamus is the crossroads of at least two known circuits. These circuits are the basolateral limbic circuit (Yakolev) and the medial limbic circuits (Papaz). Thus, if this nucleus is destroyed, there may be a significant disruption of function of these circuits and perhaps more, therefore contributing in the difficulty of new memory formation. It is now clear that this area of the thalamus may function as a data entry zone for the formation or storage of new memory. In this case, it would be the left side. Is it possible that the left hemisphere, being the more practical side of the brain, as described in chapter 2, is also responsible for the storage of new verbal memory?

Thus, one could conclude that dysfunction of a transient nature to the left dorsal medial nucleus (thalamus) would be the cause of this most curious clinical syndrome. In either case, this syndrome seems to be the result of transient disruption of this part of the thalamus on the left side.

Thus it would appear that while memory is diffusely stored throughout the cortex (outer shell), the formation of new memory is channeled through a narrow circuit (data entry zone) on its way to memory. Also, it appears that memory formation may function on the left side of the brain.

Why Is This Case of Interest to Cerebral Lateralists?

It is interesting that the left nucleus of the thalamus (not the right) is so often implicated in memory formation. I have often wondered how women seem to have such excellent memories. If new memory formation is mostly left hemisphere and the female brain is naturally blessed with higher left-sided skills (as described in chapter 2), then one should not wonder how the female remembers so well. Perhaps this is why the "absent-minded professor" is always male. Therefore, wives should not be discouraged if their husbands forget an event, for it is not lack of love but just the way the good Lord put men together. Also, because the left hemisphere is verbal/temporal and the female brain is more gifted in left hemisphere skills, one should no longer wonder why women remember dates and names better than most men. Thus, there may be a neurological explanation for the husband who forgets his wife's birthday.

It is also of interest that this area (data entry zone) is intimately related to the limbic (emotion) region of the brain. This might explain why young patients who complain of memory loss often have emotional problems, such as depression or anxiety. Their type of memory loss is always one of new memory formation. It would appear that because of the close proximity of these two structures in the brain, the lower functioning or "tone" of the emotion (limbic) area might affect the data entry zones as well. Usually, their entire neurological work-up is unremarkable except for their emotional problems. Often these patients will not improve their memory problem unless their emotional disorder is corrected.

Here we have a fine example of how intimate knowledge of brain circuitry (hardware) can help with the management of real clinical problems (software). By knowing our brains, we obtain a deeper understanding and develop a working knowledge of ourselves. We are truly "fearfully and wonderfully made."

Also, it is interesting to comment on the central location of the thalamus. You see, all incoming signals to the brain must first pass through the thalamus (visual, auditory, etc.). You might call it a signal processor. As signals pass through this body, they are then channeled to other parts of the brain. The limbic system (emotional circuits) is adjacent to the thalamus and signals go there as well as to the cognitive cortex for memory storage. You see, it was placed in this location because the Divine Engineer wanted us to be both emotional as well as intellectual. It is fascinating how we gain insight into ourselves by understanding our hardware.

Near-Death Visions (Experiences)

It is not uncommon to hear one talk of near-death experiences on television talk shows today. What are these visions like? Is there a neurophysiological explanation for this phenomenon? Can cerebral lateralization explain some of these episodes? Often we are tempted to extrapolate religion out of these stories. I believe in the concept of a divine creator. However, we might err when we use these stories to confirm or deny a religion. These visions are often of two types: (1) an encounter with persons or loved ones, and (2) a vision of a light at the end of a tunnel. Sometimes both will occur. Let's examine the first type.

Esther was a patient of mine with an intriguing story. She was a young girl who loved horses. One day while riding her horse, she had a bad fall. She struck her head severely and was in a coma for several days in the intensive care unit (ICU). When she regained consciousness, she told her interesting story. She said that while in a coma, struggling for life, she found herself in a tunnel coming to a clearing. There she discovered her father, who had been dead for five years. Her father got up and kicked her back into the tunnel and then she woke up. Was that her real father? Was she having a memory recall, an electrochemical phenomenon that causes one to totally reexperience a previous event in memory? When she had fully recovered and was able to converse with me, she explained that her father was abusive to her as a child. Although she developed epilepsy as a result of the fall, Esther went on to live a fairly normal life.

Many patients report visiting lost loved ones, others report seeing the Virgin Mary, and still others say they saw Buddha, all while near death in coma. Several questions come to mind. Did they actually visit the "real thing" or was it all an illusion? Could all the religions be correct then? This is not a book on religion but on neurophilosophy. A profound belief in God is neither foolhardy nor feebleminded. We all have the right to believe or not to believe. That is why we are given faith. I prefer to believe in a concept of God simply because of my sophisticated technical background. Anything truly sophisticated requires a sophisticated designer. Needless to say, the human brain is truly sophisticated.

The Penfield Experiments

To fully appreciate memory mechanisms and how they may apply to near-death visions, we should first discuss the findings reported by Dr. Wilder Penfield in 1959. Penfield was a neurosurgeon who performed some very innovative experiments during craniotomy (brain surgery). With the consent of the patient, microelectrical stimulating probes were inserted on and into the awake and alert human brain. The patients then described their experiences. The reason for the surgery was most often to cure focal cerebral seizures, often caused by injury or infection. Sometimes this surgery involved the complete removal of a temporal lobe. Penfield was able to map out on the brain a correlation between the area being stimulated and the type of psychical response. The temporal lobe was by far the most consistent in producing two types of psychical responses: (1) experiential hallucinations, and (2) interpretive illusions.

Experiential Hallucinations

Experiential hallucinations occurred when patients seemed to relive a previous period in their lives. Patients would see and hear the experience. In fact, not only would they see and hear the memory, but they would actually relive it "as though the stream of consciousness were flowing again as it did in the past." However, they were aware that they were actually in surgery and would always describe these responses to the doctor and never talk to the memory images. In other words, one would see one's father as he appeared years ago but would not speak directly to him. Patients were always aware of their real surroundings during the hallucination. Another interesting finding was that often the content of the memory hallucination was of very little importance in the life of the patient. Common hallucinations were of sitting with friends in the home of another friend, talking and laughing. Other patients would hear their children talking somewhere outside the room. Still others would see a familiar place at work, such as a closet where a coat would often be hung up. The

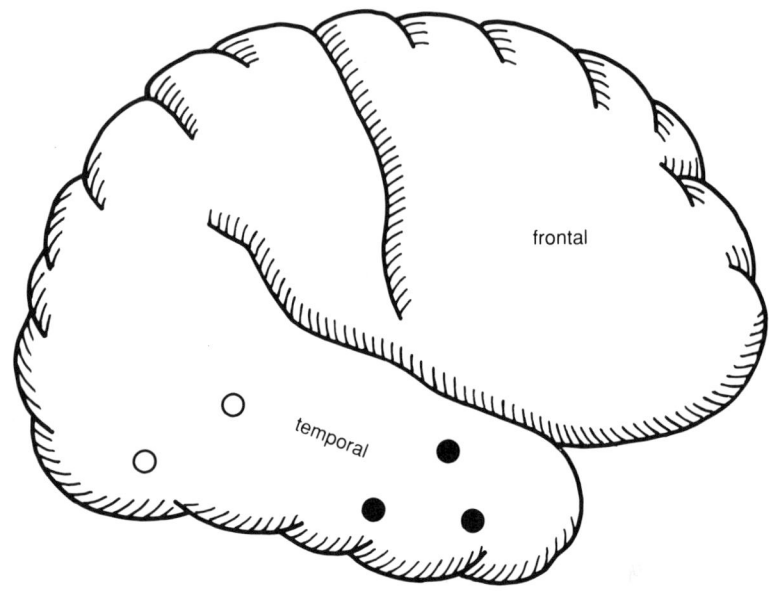

○   Interpretive Responses

●   Experiential Responses

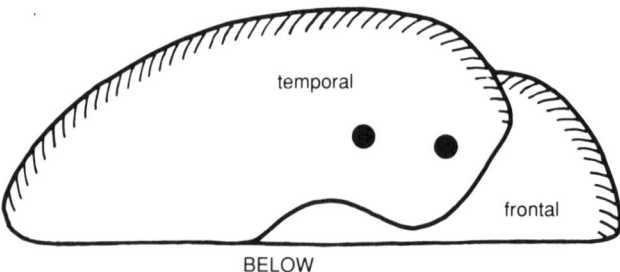

Fig. 24. Penfield's brain map: areas triggered microelectrically.

visions were very detailed and the patients would actually feel like they were there reliving the experience. The most important finding here is that we have the instant detailed reproduction of an unimportant event in the distant past life of the person involved. Thus, it was these experiments that help mold the commonly accepted theory that everything seen, heard, and felt is permanently recorded in the brain. Therefore, the reason why we so often forget is not one of inability to remember but inability to recall or access the memory cells that are already there.

Often we hear that we only use 10 percent of our brains. I do not feel that this is an accurate statement. Based on this data from Penfield, I would venture to say that we use less than 1 percent of our full mental capacity.

Interpretive Responses

Interpretive responses are subconscious conclusions about the present. These are an illusion of time and space. The patient may suddenly feel that he has had the surgery before (déjà vu). Another illusion might be that everything is strange, unreal, or unfamiliar (jamais vu). Also, the patient may feel like he is coming out of himself or leaving the world. The patient may also experience an inappropriate emotion of which the most common recorded was fear.

We see that the brain is capable of generating both the illusion of leaving the present space and recalling from memory the audio and visual information in such detail as to recreate the past if stimulated electrically on a particular side of the brain. In addition, we have learned that the area in the brain responsible for this amazing ability seems to be the temporal lobe. Thus, an electrical disturbance in this area can reproduce these psychical responses.

It is interesting to note that the temporal lobes are an extremely common area of seizure focus in the brain. Seizures are also an electrical disturbance of cerebral function. This area commonly responds in the form of an epileptic disturbance in response to metabolic changes as well as scar formation from stroke or trauma. In fact, there is an area of cortex in the mesial (inner and inferior) portion of the temporal lobe that is the most epileptogenic part of the entire brain. This area is called Ammon's horn. Ammon's horn is also a very vulnerable area of the brain for two reasons:

    1. It lies directly on the brain and is thus subject to force from brain swelling in the closed space of the cranial vault.

    2. It is supplied by small vessels that cross over the sharp edge of the tentorium (the membrane that separates the lower cerebellum from the cerebral cortex above). Therefore, these small vessels can be infarcted (obstructed) and thus deprive this area of needed blood, creating a disturbance of function. These small vessels could be vulnerable by virtue of the movement of the swollen brain and its effect on these vessels on the edge of the tentorium.

It is well known that the common end result in near-death situations is that the brain swells in its enclosed cranial vault. Having no place for the brain to expand to, it presses against the inner table of the skull. This swelling can be caused by an arrested heart, loss of blood, or respiratory failure. All these are common causes of near death.

It would not be inconceivable to think of these episodes of near-death visions as a complex psychical response secondary to electrical (epileptiform) discharges, which are a result of metabolic/vascular changes from acute brain swelling. One should not jump to religious conclusions over what very well could be a neurophysiological response.

Clinical Evidence for Seizure and Recall

Is this purely theoretical or is there any clinical evidence that a seizure focus could stimulate the cerebral cortex, resulting in an experience similar to Penfield's experiments or near-death visions?

A very pleasant and intelligent young businessman (RW) presented at my office one September morning with a most peculiar story. He was obviously embarrassed while presenting

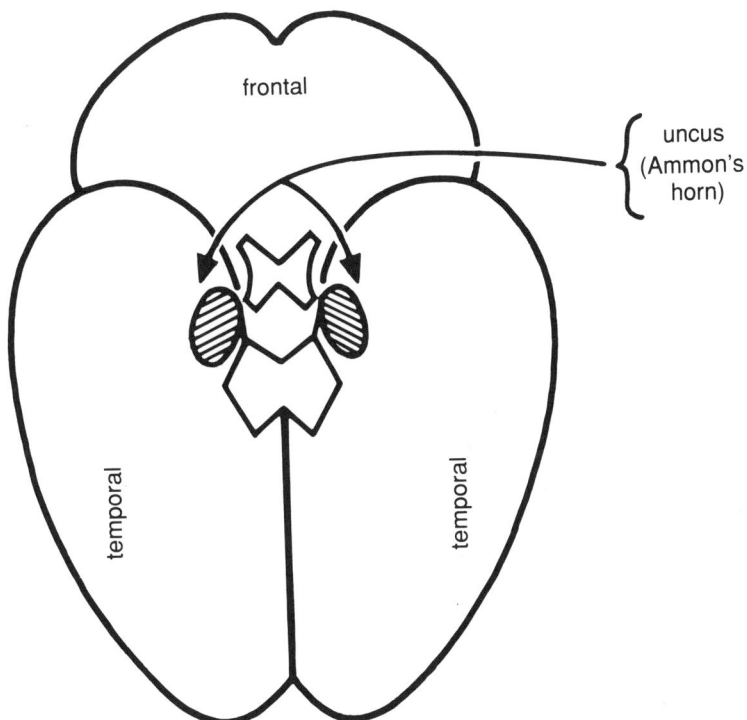

Fig. 25. View of brain from below and Ammon's horn.

his problem. His complaint was that suddenly, for no known reason, he would experience an audio/visual hallucination of cartoons. He stressed clearly that he would actually see and hear these cartoons of Bugs Bunny, Daffy Duck, etc. To be honest, my first inclination was that this person might be better served seeing a psychiatrist. However, when I questioned him further, I discovered that he would also have episodes of his past suddenly appear before him. He described these as an actual reliving of these events, even to the point of his thought content at the time. I found that these events were described in the same way by Penfield's patients over thirty years before. Now just think. If the brain is capable of storing all visual/auditory stimulation, as suggested by the Penfield experiments, there must be a lot of cartoon films stored in our brains. The natural sleep/wake cycles of children allow them to waken early on weekend mornings and watch cartoons while their parents sleep. We may not realize this, but our children must have a lot of cartoons stored in their memory cells. My patient was very concerned and wanted to know why this was happening and if there was a cure. He stated that although these episodes would only last a few minutes, they were interfering with his normal life-style.

An EEG was obtained. It showed epileptiform activity confined to the left temporal lobe. The patient was placed on the anticonvulsant Dilantin and the episodes were suppressed. It is interesting to note that a follow-up EEG, while the patient was well controlled on seizure medicine, did not show the epileptiform activity. We, therefore, have clinical evidence for a seizure focus stimulating the temporal lobe, with results similar to the Penfield experiments.

There appears to be evidence that near-death visions might very well be visual-auditory hallucinations triggered by focal epileptiform discharges, confined to a disfunctioning temporal lobe. Perhaps this is also the cause of the "my life passed before my eyes" phenomenon, which has also been described in near-death experiences. One may conclude that electrical discharges in the temporal lobe may trigger a rapid visual movie, previously stored in the brain.

Light at the End of a Tunnel

This is another common experience in near death. Often the person describes darkness all around with a light that appears off in the distance. Many times it varies in size, suggesting movement toward or away from the source. Does this also have a neurophysiological explanation?

As we discovered in chapter 1, the part of our brain that is used for vision is the occipital lobe (rear part of the brain). The vascular supply of this area is unique. The calcarine cortex is the part of the occipital lobe responsible for central vision. This area is remarkable for the fact that it is well nourished, having two separate blood supplies (posterior cerebral artery and middle cerebral artery). Thus, one could conclude that as the blood supply is cut off during near death, this area would persist in function longer, causing the impression of a central object or light. As the blood supply varies to this area, so does the relative function of central vision to peripheral vision, thus producing the illusion of coming to or away from the end of the tunnel.

This type of experience would therefore have a vascular cause rather than an electrical one, as previously described. Thus, there are two important neurophysiologic explanations for these experiences, electrical and vascular. This, however, should not discourage us from a belief in God. I strongly believe that we and nature are the product of divine engineering. However, we should be careful not to make a religion out of near-death experiences.

Storage and Retrieval

How does the brain store data in memory? And when it is in memory, how is it recalled? Are different types of memory stored differently? To fully appreciate the mechanisms of memory formation and retrieval, we must first understand the difference between *retrograde* and *anterograde* amnesia (retrieval and storage defects, respectively).

Anterograde amnesia is the inability to recall events occurring *after* a traumatic cerebral event. It is a disorder of new memory formation. Events prior to the traumatic event, i.e., childhood, schooldays, etc., are relatively well preserved. A good case is Wernicke's amnesia, mentioned previously in this chapter.

Retrograde amnesia is the inability to recall events or memory *prior to* a traumatic cerebral event. This person will have no difficulty recalling present-day events. He is the same as you or I in regard to new memory formation. However, he may know nothing of his past. He may not know his parents or wife or boss. He may not even know his name. He has to relearn himself with the help of others. Little is known about retrograde amnesia, and there is no specific therapy or drug. These people are a stranger to themselves and at first appear lost. Often there are no neurofocal problems. They are fully alert, walking and talking, but must totally relearn their world. This syndrome is very rare and often it is the result of a blow to the head (closed head injury).

Man a Religious Animal: Is Religious Memory Processed Differently?

As described previously, one should not come to religious conclusions on the grounds of what might be a normal neurophysiological mechanism of memory reproduction, such as those experienced in near-death visions. However, there may be evidence for divinity and the brain. Are there differences in storage among the various types of information? For example, as any anthropologist or archeologist will tell you, man is a religious animal. This is obviously not new. Is it possible that man, being a religious animal, may store religious information differently from other information? Would this not again fit our concept of cerebral lateralization of behavior following skill? If religious information is processed differently, it must be for a reason. Could the reason be that we *are* the product of divine engineering and a special area of the brain was designed for man's relationship with his maker (the Divine Engineer)? I will explain, using two interesting case histories of patients of mine.

Case 1. PL is a very pleasant man, who presented with a very interesting story. He works installing garage doors. One day while walking about his truck, he bumped the top of his head on a steel beam. He did not lose consciousness, but the injury did require stitches. This was done and PL returned to work an hour later. A few hours after that he was found lying near his truck unresponsive. He was taken to the hospital, which was where I first met him. After a few hours, PL became fully responsive but had no knowledge of his own personal information or any world events. It was as if he had been dropped on this planet from a spacecraft. He did not know he was married or who his wife was. He did not know who was the president of the United States, or any president for that matter. He could not even name the first president. He could not name the capital cities of any of the states. He did not know his parents, his line of work, or where he came from. However, he had intact recall of all of our meetings from day to day, so his memory formation was still good. Hence, no anterograde amnesia, only pure retrograde.

What might be of interest (when compared with the second case later on) is that this patient did not have a significant religious past. He did not attend either church or synagogue, and his parents were not religious. When questioned about his religious knowledge, PL responded that he knew there was a difference between Christianity, Judaism, and Islam, but he was not sure what it was.

As the weeks passed PL tried to capture his past by drawing pictures of visions that would come to mind. These are actual drawings provided by the patient. One could construe that being male and therefore right hemispheric in thought this visual/spatial method of attempting to grasp his past should not be too surprising. PL was able to draw road maps of where he lived with good accuracy. One particular drawing I found most relevant. He drew a picture of a cliff, which appeared to be by the sea. His wife stated that this picture was almost identical to a Valentine's Day card made by him and given to her on their first Valentine's Day together. Another was a series of narrow stairs going down through a hole in the floor, much like aboard ship. The patient had no idea what the drawings meant, but his wife informed me that he had been in the navy and spent a great deal of time aboard

Fig. 26A. Drawing by patient in case 1 of a cliff by the sea.

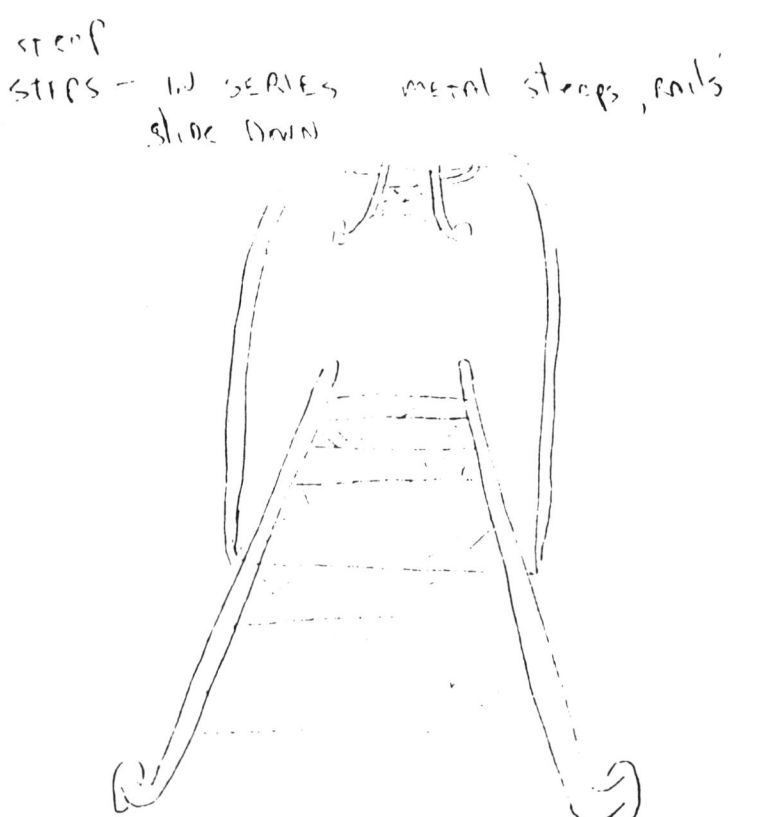

Fig. 26B. Drawing by patient in case 1 of a series of narrow stairs going down through a hole in the floor.

an aircraft carrier. It appears that he was searching for his past through his right (visual/spatial) hemisphere.

It is interesting to note that PL's MRI of the brain was normal, as was his EEG. However, neuropsychiatric testing indicated a left temporal lobe lesion. Is it possible that memory recall is controlled from the temporal lobe, the left temporal lobe to be more precise? This would be consistent with the Penfield experiments described previously. It is interesting to note that PL did not dream. However, after treating him with Hydergene (a medicine sometimes helpful in patients with dementia), PL began to dream and to develop some recall of his past. Could there be a relation between dreaming and memory recall? I think so.

Case 2. JK also developed a sudden onset of retrograde amnesia in a very similar manner. He too drove a truck for a living and struck the top of his head on a steel pipe while getting up from a bent position underneath his truck. Afterward he had no idea where he came from or what he was doing there. For that reason, JK was taken to the hospital where I first met him.

JK's pattern of memory loss was also very similar to our first case. He had no knowledge of any presidents of the United States or the capitals of any states. He did not know who his parents were or where he had come from. Otherwise, he had a normal neurological exam.

It is also interesting to note that the type of blow inflicted on both patients was the same. The force was directed from the top downward. It is a well-established neuropathological fact that this pattern of injury is often found with lesions (contusions or bruises) to the mesial portions (inner) of the temporal lobes. This might be a key to the functional localization of memory recall in the human brain.

After talking with JK's mother, I learned that our patient had become a born-again Christian about six months previous to the incident. This is a most interesting characteristic because when the patient was questioned about religion, he presented significant knowledge about Christianity and acted as if attempting to convert me.

Is it therefore possible that religious information is stored differently than other types of information? Is it possible then that there might be a localized religious information area in our brains not yet discovered? Many years ago when physicians found a body part for which there was no obvious function, they called it an "appendage," a leftover from a previously functioning body part (phylogenically) from our evolutionary ancestors. After all, we take out the appendix and the patient lives just as well without it. You can take out the spleen and the patient goes on living. However, with the advent of technology we have long since realized that there just might be no true appendages and that all the organs have a specific function filling a biochemical need. Therefore, why could it not be true for the human brain as well? For if it is possible that there is a specialized religious area in the human brain and everything has a legitimate reason, then we are truly religious animals and God may, indeed, exist in whatever form you conceive him. Perhaps this religious center is there for that reason; it is man's information link with his maker. Truly, there is so much more to learn about our brains. We are not even scratching the tip of the iceberg for we are indeed "fearfully and wonderfully made." Considering the similarity of the blow inflicted on both of our patients and the relative probability of the involved areas being confined to the mesiotemporal lobes and its well-established high degree of epileptogenicity (tendency to cause seizure activity), one could not be too surprised to find near-death religious visions as described by Esther earlier in this chapter. Is the mesiotemporal lobe the religious center of the human brain? It could be. . . .

Summary

We have learned the two basic types of memory mechanisms used, storage and retrieval. We have also learned that defects in these mechanisms may cause amnesia, anterograde and retrograde, respectively. We have learned that these mechanisms may be localized in different areas of the brain. We see that data entry may be a lateralized function in the thalamus and that recall of old memory may be the result of lateralized functions in the temporal lobes. Finally, we discussed the possibility that man, being a religious animal, may process religious information differently. Does this again illustrate the concept of behavior following skill? Does this indicate the existence of a divinity? That will be left to you to answer. Spirituality is individual.

8. Conclusion

How does knowledge of cerebral lateralization help us in our relationships with others? Can our children benefit if we know more about how our brain functions? Should educators be more aware of the importance of nonverbal (right hemisphere) skills in early childhood development?

Lateralization Understanding and Tolerance

Romance and Lateralization

We have learned in the previous chapters that there are differences between the sexes. We have also learned that these differences are not ones of superiority versus inferiority but rather lateralization (side to side), a difference through equality. We have also learned that just because one is right-handed or left-handed does not necessarily mean that one will be endowed with left or right hemisphere skills. By the same token, any female can be just as right hemispheric in skill as any male and any male as left hemispheric as any female for there are vertical (global high-functioning) differences as well as lateral. One should not confuse statistics from a large group with the single individual. Logic does not always flow in both directions. Remember, *All squares are parallelograms, but not all parallelograms are squares.* Thus, there is no room for prejudice of the individual. However, there are differences between the sexes and it might be that this difference is what attracts us to each other. Therefore, let us not be so concerned with changing our lover or spouse but rather recognize these differences and have fun with them. What can we gain from this knowledge and understanding that will help us in our sexual relationships?

Since puberty, I have been fascinated by the attraction between the sexes. I have believed that this attraction went beyond the physical to one of cerebral magnetism. Our discoveries in the preceding chapters bear out this cerebral magnetism through the differences. How often we have heard that there is that "certain something" in the other person that we can't identify that attracts us: the way he/she walks, the way he/she talks, etc. There is no question that the physical aspects are of tremendous importance. However, there is that "touch" of the opposite sex that somehow makes us complete, that tone of voice or attitude toward events that draws us together. We have all seen the bachelor pad that contains no curtains and only an empty refrigerator, except for the beer. Needless to say, there are things missing there that almost any female brain will detect. However, his car will probably be in great shape. Again, we come to the inner and outer sphere concept or these differences, the hub and the tire as described in chapter 2. Thus romance becomes not just two people getting together but two halves becoming whole, a completeness that may potentially provide the proper and holistic environment for offspring. Thus, the children are afforded all possible opportunities to be successful and independent in their adult lives. There is great beauty in this design, and I do believe it was a design; it is too wonderful to be just chance. I'm not being religious but rather spiritual. We are "fearfully and wonderfully made," said King David. I believe that this goes beyond the physical body and encompasses the entire design of our survivability, which of course includes romance. All can appreciate this beauty, regardless of one's religion.

What can we learn from this that may help our sexual relationships? Often we search for the perfect mate, the person who meets our needs and desires to perfection. Sometimes, during dating, there is a process of molding. We desire to shape the other person to more perfectly fit the ideal that we have in mind. This could be in the form of hobbies, personalities, or even the variety of sexual needs one may have. Sometimes we believe that those few remaining traits that irritate us will, through love, be corrected after marriage. Unfortunately, many times it is only discovered that neither love nor marriage solves the problem, and the spouse continues to be irritated. Sometimes this irritation grows to the point where the only solution is divorce. Love is great, but far too often it is not enough to make the changes. Perhaps some of these differences are ingrained or formed in the womb prior to birth and thus can never change for love or money. Perhaps this should be taken into consideration prior to marriage. However, there is another kind of difference, that which was not clearly detected prior to matrimony, only to present itself later. "It didn't bother me before, but now I can't stand it," or "It didn't bother me prior to the marriage, but now I just want to and he/she still won't let me." Sometimes the molding process never stops. "Why couldn't he/she be more like him/her?" "Why can't he/she be more like me?"

Another interesting observation is the unique and amazing ability that women seem to have in recalling, verbatim, past conversations or events and their dates. Needless to say, this often has been the cause of many arguments between lovers. "He didn't remember and so he just does not care." It could be very possible that because of the relative superior functioning of the left hemisphere (verbal/temporal) in the female brain compared to the mainly right hemispheric (visual/spatial) male that she would remember while he would not. Therefore, it should not be taken personally. The good Lord simply did not endow as many men with superior left hemisphere skills as women. Thus, it is only natural that she would remember while he did not; it is not simply the result of lack of love. One of the reasons I had in mind for writing this book was the hope that, with understanding, we could learn to accept each other. Perhaps a marriage or two might be saved because of this writing. We should learn about our differences and accept them. I do not believe that marriage is simply the coming together of two bodies but rather two minds. Thus, it is holistic and complete. The physical is important, but there is that something the other sex provides that we lack. After all, there are only two sexes and two hemispheres to the brain. Thus, it should not be too surprising that romance might be a coming together of two hemispheres to achieve the goal of completeness.

The concept of sexual cerebral lateralization teaches us that often these differences are determined prior to birth and are therefore unlikely to change. Also, sometimes it is the difference that attracts us in the first place. Now some might ask why then should we want someone to change? Perhaps it is that endless search for the perfect someone, the ideal mate, the 36-24-34 rocket scientist nymphomaniac or the six-foot, six-inch tall, dark, and handsome, tender-hearted millionaire. However, life is not perfect and perhaps tolerance is sometimes the better part of visionary gratification. Needless to say, communication is the most important solution to most of these problems. However, there might be differences that we just might have to accept if we want to stay together.

The Irritating Difference

What differences are there between the sexes that might be less likely to change and present themselves as a constant "thorn in the side"? What are those differences that might have been hardwired in in utero and thus resistant to change?

One thing that might help is understanding the animals. Often a little knowledge of the primates may help us understand ourselves. Perhaps this is one of the reasons why I, as well as millions of others, have been such a dog lover. I am fascinated by the subtle yet definite differences between the sexes of dogs and even horses, as reported to me by other lovers of animals and their veterinarians. In chapter 2, we discussed how male dogs will often roam about the neighborhood only to return home when hungry, and how female dogs makes excellent companions because of their tendency to stay close to home. This is the reason my mother always insisted on female dogs. I remember one of our males, Rebel, had to be given away simply because he was too difficult to train and often got away.

Is there a lesson here that might help us in understanding our sexual partners and assist in preserving our relationships with the opposite sex? Sometimes there are demands placed on the spouse to be more like his or her mate. We should not be afraid of the differences but rather accept them. After all, our differences are not only what separates us but what attracts us as well. Perhaps we should give our mates space to express their gender individuality. Maybe we should not be alarmed because our spouse might desire to spend a day or so with their gender friends. We should not try to demasculinize or defemininize our husbands and wives, which might put them in such a position that they can't feel comfortable. Sometimes these differences are hardwired in at birth, and we're just trying to push square pegs through round holes. Perhaps if we give our spouses the freedom they need to grow in their respective genders, we might find ourselves all the more attracted to them. Females become more female and the males more male, and the attraction grows. Sometimes we must give before we receive.

Let me explain. I enjoy reading with my cocker spaniel, Maggie, in my lap. I learned long ago that if I held her there so she could not leave when she wanted, her desire to sit on my lap was greatly reduced. However, if I did not restrain her, she was more apt to spontaneously sit there. She felt secure knowing that if she needed to leave for any reason, a drink of water or to look out the window, she could. I felt secure knowing that Maggie would be more apt to return to my lap when she was done. A little freedom may go a long way in return. If one is too jealous or possessive, one may lose in the long run. Remember the square pegs and round holes; we should appreciate and not hide our differences. Viva la difference!

Child Development and Lateralization

Much of the differences, even in young growing brains, are determined prior in utero. This is not to belittle the importance of environmental influence but simply to show that sometimes these differences are hardwired in and are not necessarily the result of good or bad events in the child's life. There is a whole myriad of skills representative of one of the hemispheres, and a child may be endowed with special talents and an interest in a specific field. This is not to say that a child is born to be a mechanical engineer or an accomplished pianist, but if born with the typical right hemisphere orientation and adequate parental guidance, he could be what he desires to be in one of those right hemisphere professions in which he is skilled. The purpose of the parent is not to dictate the child's professional future but rather to see that the child is properly exposed to the various fields and have the child discover their own skills and interests. Exposure and opportunity, this is the true function of parenthood. Perhaps this is the reason the good Lord decided that it was only sex between male and female that was to be blessed with offspring, thus allowing the child exposure to both hemispheres, the father providing one side, the mother providing the other. Thus, the child is afforded a balanced upbringing. This is not to say that homosexual parents are always deprived of this for they often become attached to opposite personalities, thus affording some balance for offspring as well.

We have also learned that there are no late bloomers but rather undiscovered talents. In telling my own personal story about how I dealt with dyslexia, it was not until high school that my teachers began changing their view of me. I do not believe it was because somehow a burst of cerebral skills were magically bestowed on me during that time but rather a previously developed skill was allowed to be exposed. There is no doubt that this skill was determined at birth. Thus, it is the responsibility of adults to see that adequate exposure and opportunities are afforded our youth. We have all heard that "a mind is a terrible thing to waste." We should therefore be more concerned with giving children exposure and opportunity rather than classifying them.

To make this more clear, we should understand that the first several years of education deal with stressing primarily the left hemisphere (reading, writing, arithmetic), and it is not until later that the right hemisphere begins to be stressed (geometry, physics, precalculus). Educators have learned to be patient with those who show evidence of relative hypofunctioning of the left hemisphere (dyslexia, hyperactivity), while cerebral lateralization suggests they could very well be blessed with special right hemisphere skills. You see, again there is a sort of difference through equality here, a natural tendency for balance. Thus, parents should not be too alarmed because their child is not "up to par" in the early stages of school. We should never say the he/she is just slow. Prejudicing a child like that at such an early stage serves no purpose and can only do harm. Remember, as stated previously, there are no late bloomers, just undiscovered talents. Love and encouragement must still continue. We all know Einstein did not reveal his genius until much later in adulthood.

We have also learned that the left hemisphere seems to mature later than the right, and that the male brain matures after the female. Therefore, the typically right hemispheric male may be at a disadvantage in the early stages of education. For that reason we should not be hesitant to keep some of these students at home a year or two so that they are not prejudiced against early on. As stated previously, any female could also be right hemispheric in skill orientation; thus, they should not be ignored as well. However, the general trend indicates that the male is more likely to be plagued with this problem.

We have also discussed the concept of genius and subconscious logic. Could many of these skills originate from the nonverbal right hemisphere? The brain is far too complex to simply measure its skills in a simple three-hour test. How do you measure creativity and inspiration? The child needs ample exposure to find hidden talents, those islets of genius in us all. Exposure can be in the form of toys, museum trips, and various adult consultations. We have also discussed the possibility that some of these geniuslike skills might be potentially developed in any one of us. This was brought out by the discovery of the graduate student (chapter 5) who taught himself to become a "calendar brain." The answers just seemed to pop into his head. Is this a form of inspiration or intuition, and if so, should this be taken into consideration in the education of our youth? If inspiration originates from the nonverbal right hemisphere, then perhaps right hemisphere skills should be encouraged in our children. Should we not be encouraging creativity and independent thought in all forms of education (music, literature, arts and sciences)? Should we not encourage our children to draw pictures of scenes not yet visualized and music not yet heard? We readily accept the obvious verbal world we live in; however, an entire right side of our brain is nonverbal; that should be developed as well. Surely creativity and inspiration are a very vital component of our culture as well as our economy. Without ideas there is no art. Without ideas there is no product for business to produce. Many of these ideas come from the nonverbal right hemisphere, for no single hemisphere has total control of creativity. Perhaps we can't afford not to take seriously this aspect of human talent.

Are You a Left Brain or a Right Brain Person?

This is not to say that there are only two kinds of people. In chapter 6, when we talked of the skewing of lateralization, we saw that this relationship of skills and behavior between the left and right, sensitivity and motor, could cause a myriad of combinations. However, often we hear someone say, "You must be a left brain person," or "You must be a right brain person." What is meant by this? Are there special characteristics that can help us identify one as a right or left brain person?

The Left Brain Person

If you are female, it is typical to be a left brain person. If male, you most probably are quite global in your cerebral talents, using both hemispheres well, a superbrain. You like to read novels or history. You might appreciate poetry. You might be sensitive of what people say about you. You probably did fairly well in the first several years of school. You might have done well in math but did not like calculus. When receiving directions, you write them down in longhand.

When listening to music, you are more apt to listen to the words. Although you might actually play piano well, you have never really seriously written music. You were never very serious about buying an expensive stereo system. The $150 set works rather well and you see no real reason to change. Besides, you probably still have your eight-track tapes. However, you are concerned about the arts.

You most likely do not subscribe to *Car and Driver.* You are more apt to recognize cars by their size, color, cleanliness, and amount of chrome. You prefer someone else to do the driving. You are more concerned about getting lost than your right hemisphere friends. You have someone else fix your car. Besides, you prefer not to get dirty anyway.

You are careful about the words you select when talking to others. You are a fairly good speller. You tend to remember names and dates, but you may forget numbers. In fact, you might have some favorite names yourself. You would be more apt to write letters than your right hemisphere friends. You also enjoy staying in contact over the phone.

You consider yourself (and may actually be) rather astute at "reading" the true intentions of others. You are fairly observant of your surroundings (especially colors). You might consider yourself (and actually be) a good detective. You probably do not have to ask people to repeat themselves twice.

You've always wanted to and probably have mastered a foreign language. Although you might enjoy traveling, you hate living out of a suitcase. You probably know the difference between a predicate nominative and a past participle. However, you can't tell the difference between a muffler bearing and a rocker arm.

Careers you might have considered include law, foreign languages, teacher, history professor, novelist, playwright, psychiatrist, medical doctor, interior decoration, sociology, public relations, journalist, clothing design, politics, criminal investigation, concert pianist.

The Right Brain Person

If male, you are typical. If female, you are probably global in your cerebral talents, using both hemispheres well, a superbrain. You are probably not a big reader of novels, perhaps preferring movies. Although desirous of success in life, you are probably not very concerned about what people say about you. The first several years of school might have been fairly hard for you; you probably did better in the latter part of high school. You might have been hyperactive as a child. You probably found calculus interesting. When receiving or giving directions, you write a little map.

You enjoy the melody of music much more than the words. Although receiving little formal musical education, you play an instrument and have written some songs yourself. You are more apt to play by ear. You have probably always wanted to buy an expensive stereo system. You threw out your eight-tracks long ago and now have a compact disc system.

You are probably well acquainted with the various automobile magazines. You know cars, model and year, by the shape of their designs. You enjoy driving and are not afraid of getting lost. Besides, you like making decisions. You would prefer, if possible, to work on your own car. Getting dirty does not bother you, if temporary.

You are not very careful about the words you select when talking to others. Your left hemisphere friends consider you at times careless in your words, but you don't agree. Besides, words aren't important to you. You can't spell. Spelling did not interest you anyway. You have difficulty remembering names and dates but consider yourself good (and probably are) with numbers. You hate writing letters and tend to not stay in contact when friends move out of town. You ask people to repeat themselves: "Come again," "What's that?"

You hated French, don't know a past participle from a predicate nominative, and could not care less. However, you do know that muffler bearings don't exist. For relaxation, you consider building models or solving differential equations. Navigation, machines, electronics, and physics might fascinate you. You consider yourself practical, a nuts-and-bolts kind of person.

Careers you might have considered include architecture, engineering, physics, math professor, mechanic, computer designer or programmer, surgeon, music composer.

How about You and the Ocular Dominance Test?

Unfortunately, there are no conclusive studies about the Ocular Dominance Test and whether one is right or left hemisphere in mentation. However, studies are needed and would be very interesting. One might think, however, that if one were left eye dominant, one would be more apt to be right hemispheric in cerebral function and if right eye dominant, left hemispheric. Remember the dyslexia patients who were able to read better through a mirror, reading right to left, thus leading with their right hemisphere (Orton, 1925). As we have learned, dyslexic persons are apt to be blessed with right hemisphere skills. Try the test yourself. Perhaps you will learn something about your brain. It is my hope that by including an Ocular Dominance Test with this book, needed research in the relationship between eye dominance and lateralized skills will be started.

Summary

We now see that cerebral lateralization looks at the brain from the point of view of hardware as opposed to software. The brain is hardwired together at an early stage in life and much is determined prior to birth. This process of cerebral development (hardwiring) causes a difference between the sexes, a difference in sexual preference, and a difference in type and degree of skills among people. This difference may also affect behavior as well as personalities, and that behavior may follow skills endowed prior to birth. We have also learned that there may be a relative law of conservation between the hemispheres. Some may have right hemisphere skills, while others have left hemisphere skills. The difference may not be one of superiority or inferiority but rather lateralization (side to side). Therefore, we are all different but equal. Should we not learn to accept different people with different needs?

Cerebral lateralization is more than just left hemisphere and right hemisphere skills. It is the view that looks at the brain from the point of localized areas of specific functions (memory and learning) all working harmoniously together to form the individual. The brain is a marvelous machine with various parts. When one looks at the brain this way, one wonders where the soul or individual lies. If our brain is composed of parts, then where is the owner of these parts? As mentioned earlier, the word *individual* comes from the Latin *individuus,* meaning "can no longer be further divided." Perhaps our forefathers also understood this mechanistic approach to the brain.

So, Why the Attraction between the Sexes?

Learning about God through studying His handiwork . . .

God said, let us make man in our image . . .
So God made man fall into a deep sleep. And while he slept, he took one of his ribs and enclosed it in flesh. God build the rib he had taken from the man into a woman and brought her to the man. The man exclaimed, This at last is bone from my bones and flesh from my flesh! This is to be called Woman, for this was taken from man.
(Gen. 1:26, 2:21–23)

We must realize that the good Lord was attempting to explain the origin and, in turn, the difference between the sexes to slaves. These were not educated people. They believed that the origin of thought was from the heart, just as the ancient Babylonians thought it to be from the liver.

We know that there are two halves to the human brain, divided into skills unique to each hemisphere. Thus, it would not be too difficult to comprehend that a sex might represent a hemisphere; after all, there are only two hemispheres and two sexes, and scientific data seems to be pointing in this direction.

A college degree in physics is not needed to know that opposite charges attract. This attraction helps satisfy a deficit of charge within the pole. A positive pole lacks a negative charge and is therefore attracted to it and visa-versa. Narrow hips are attracted to slightly larger ones, and soft skin is attracted to a more muscular build. Deep voices are attracted to soft higher ones.

Cerebral lateralization teaches that this attraction is cerebral in origin and not just physical, an attraction between the hemispheres, a desire to become whole again. Thus, it is this union of hemispheres that is blessed by offspring, providing a whole and balanced environment for development.

The ancient Hebrews, being unaware of the brain as the origin of human thought, believed intellect emanated from the heart inside the chest. For this reason, Moses and the good Lord used a rib to explain a cerebral origin of the sexes to the Hebrew slaves; that Adam was once whole, neither male nor female; that half of his mentation was taken to form the female sex. Adam recognized this cerebral transfer and thus said, "Bone from my bone and flesh from my flesh." Seeing that Adam was made "in God's image," perhaps God is neither male nor female but a balanced mixture of both, comprising the supportive characteristics from the male and the nurturing aspects from the female. This was the first lecture on cerebral lateralization in recorded history.

We truly are functionally equivalent, different but equal. This is the art of neurophilosophy.

Bibliography

Baer, D. and P. Fedio. "Quantitative Analysis of Interictal Behavior in Temporal Lobe Epilepsy." *Arch Neurol* 34 (August 1977): 454–67.
Barnes, D. "Bird Brain Switch Leads to New Song." *Science* 241: 1434–435.
Boklage, C. E. "Schizophrenia, Brain Asym. Development and Twinning, Cellular Relationship with Etiological and Prognostic Implications." *Biol. Psychiat.* 12 (1977):19–35.
Bronson, F. H. and C. Desjardins. "Aggression in Adult Mice: Modification by Neonatal Injections of Gonadal Hormones." *Science* 161 (1968): 705–6.
Buffery, A. H. and J. A. Gray. "Sex Differences in the Development of Spatial and Linguistics Skills" in *Gender Differences: Their Ontogeny and Significance,* eds. C. Dunsted and D. C. Taylor. New York: Churchill Livingston Inc., 1972.
Caplan. "TGA: Criteria and Classification." *Neurology* 36 (March 1986): 441.
Charness, N., J. Clifton, and L. MacDonald. "A Case Study of a Musical Mono-Savant: A Cognitive Psychological Focus." *The Exceptional Brain: Neuropsychology of Talent and Special Abilities,* eds. L. K. Ubler and D. A. Fein. New York: Guildford Press, 1988.
Diamond, M. C., P. E. Johnson, and C. A. Ingham. "Morphologic Changes in the Young, Adult and Aging Rat Cerebral Cortex, Hippocampus and Diencephalon." *Behav Biolo,* 14 (1975): 153–174.
Diamond, M. C. "Age, Sex, and Environmental Influences." In *Cerebral Dominance: The Biological Foundation,* eds. Geschwind and A. M. Galaburda. Cambridge, Mass.: Harvard University Press, 1984.
Diamond, M. C., G. A. Dowling, and R. E. Johnson. "Morphological Asymmetry in Male and Female Rats." *Exp Neurol* 71 (1981): 261–68.
Dohler, K. D. "Is Female Sexual Differentiation Hormone Mediated?" *TINS* 1 (1978): 138–40.
Dorner, G., B. Schenk, and B. Schmiedel. "Stressful Events in Prenatal Life of Bi and Homosexual Men." *Exp Clin Endocrinol* 81 (1983): 83–87.
Dorner, G., F. Gotz, and W. D. Docke. "Prevention of Demasculinization and Feminization of the Brain in Prenatally Stressed Male Rats by Prenatal Androgen Treatment." *Exp Clin Endocrinol* 81 (1983): 88–90.
Edwards, D. A. "Mice: Fighting by Neonatally Androgenized Females." *Science* 161 (1964): 1027–28.
Eidelberg, D. and A. Galaburda. "Inferior Parietal Lobe, Divergent Architectonic Asymmetries in the Human Brain." *Arch Neurol* 41 (August 1984): 843–52.

Fishbein, W. and F. Jidong. "Sexual Dimorphism of Sleep in the Rat." Doctoral program in neurocognition, the City College and Graduate School, City University of New York.

Fisher, C. M. and R. Adams. "TGA. Transamer" *Neurolo Ass* 83 (1958): 143–46.

Gesell, A. and L. E. Ames. "The Development of Handedness." *J Genet Psychol* 70 (1947): 155–75.

Goldman, F. S. "Neuronal Plasticity in Primates Telencephalon: Anomalous Projections Induced by Prenatal Removal of Frontal Cortex." *Science* 202 (1978): 768–70.

Gordon, H. W. "Cognitive Asymmetry in Dyslexic Families." *Neuropsychologia* 18 (1980): 645–56.

Gorlick, P. "TGA and Thalamic Infarction." *Neurology* 38 (March 1988): 496.

Haseltine, F. P. and S. Ohno. "Mechanisms of Gonadal Differentiation." *Science* 211 (1981): 1272–278.

Heir, D. and W. F. Crowley, Jr. "Spatial Ability in Androgen-Deficient Men." *N Engl J Med* 306 (1882): 1202–205.

Hinge, H. O. "The Prognosis of TGA." *Arch Neurol* 43 (July 1986): 673–76.

Hooker, E. "The Adjustment of the Male Overt Homosexual." *J Prejective Techniques* 21, 1 (1957) 18–31.

Hynd, George W. and Semrud-Clikeman. "Brain Morphology in Developmental Dyslexia and Attention Deficit Disorder/Hyperactivity." *Arch Neurol* 47, 8: 919–26.

Kanner, L. "Autistic Disturbances of Affective." *Nerv Child* 2 (1943): 217–50.

Kushner, M. "TGA: A Case Control Study." *Ann Neurol* 18 (1985): 684–91.

Landwehrmeyer, B. and J. Gerling. "Patterns of Task-Related Slow Brain Potentials in Dyslexia." *Arch Neurol* 47 (July 1990): 791–97.

Lee, G. F. and D. W. Loring. "Severe Behavioral Complications following Intracarotid Sodium Amobarbital Injection, Implications for Hemispheric Asymmetry of Emotion." *Neurology* 38 (1988): 1233–236.

LeVay, Simon. *Science* (August 1991).

McKusick, V. A. "Mapping and Sequencing the Human Genome." *N Engl J Med* (April 6, 1989): 910–15.

McKusick, V. A. "Mendelian Inheritance in Man." Catalogs of Autosomal Dom., Autosomal Res., and X-Linked Phenotypes, 8th ed. Baltimore: Johns Hopkins University Press, 1988.

McLean, J. M. and F. M. Ciurzak. "Bimanual Dexterity in Major League Baseball Players: A Statistical Study." *N Engl J Med* 2 (1982): 1278–279.

Mittwoch, V. "Lateral Asymmetry and Gonadal Differentiation." *Lancet* 1 (1975): 401–2.

Money, J. and A. A. Ehrhardt. "Gender Dimorphic Behavior and Fetal Sex Hormones," in *Recent Progress in Hormone Research,* Vol. 28, ed. E. B. Astwood. New York: Academic Press Inc., 1972.

Nass, R., S. Baker, and P. Speiser. "Hormones and Handedness; Left-Hand Bias in Female Congenital Adrenal Hyperplasia Patients." *Neurology* 37 (1987): 711–15.

Oldfield, F. O. "Handedness in Musicians." *Br J Psychol* 60 (1969): 91–99.

Oliveirea, V., J. M. Ferro, and J. Foreid. "Kluver-Bucy Syndrome in Systemic Lupus Erythematosus." Dept of Neurol., Hospital de Santa Maria, 1600 Lisbon. *PRT-J Neurol* 236, 1 (1989): 55–56.

Orton, S. T. " 'Word Blindness' in School Children." *Arch Neurol* 14 (1925): 581–615.

Penfield, W. and L. Roberts. *Speech and Brain Mechanisms.* Princeton, New Jersey: Princeton University Press, 1959.

Peterson, J. M. "Left-Handedness: Differences between Student Artist and Scientist." *Percept Motor Skills* 48 (1979): 961–62.

Peterson, J. M. and L. M. Lansky. "Left-Handedness among Architects: Some Facts and Speculations." *Percept Motor Skills* 48 (1979): 961–62.

Raisman, G. and P. Field. "Sexual Dimorphism in the Preoptic Area of the Rat." *Science* 173 (1971): 731–33.

Ranjan, Duara, M. D. and Alex Kushch. "Neuroanatomic Differences between Dyslexic and Normal Readers on Magnetic Responance Imagining Scans." *Arch Neurol* 48 (April 1991).

Rimland, B. "Savant Capabilities of Autistic Children and Their Cognative Implications," in *Cognitive Defects in the Development of Mental Illness,* ed., G. Serban. New York: Brunner/Mazel, 1978.

Rimland, B. "Inside the Mind of the Autistic Savant." *Psychology Today* 12, 3 (August 1978): 68–80.

Ross, E. D. "Modulation of Affect and Nonverbal Communication by the Right Hemisphere," in *Principles of Behavioral Neurology,* ed. M. M. Mesulam. Philadelphia: F A Davis, 1985.

Saghir, M. T., E. Robins, and B. Walbran. "Homosexuality, Sexual Behavior of the Male Homosexual." *Archives of General Psychiatry* 21 (1969): 219–229.

Sano, F. "James Henry Pullen: The Genius of Earlswood." *J Ment Sci* 64 (1918): 251–267.

Schwartz, G. E., F. J. Davidson, and F. Maer. "Right Hemisphere Lateralization for Emotion in the Human Brain: Interaction with Cognition." *Science* 19 (1975): 236–88.

Speedie, L. and K. Heilman. "Amnesic Disturbance following Infarction of the Left Dorsal Medial Nucleus of the Thalamus." *Neuropsychologia* 20, 5 (1982): 597–604.

Squire, L. R. and R. Y. Moore. "Dorsal Thalamic Lesion in a Noted Case of Human Memory Dysfunction." *Ann Neurol* 6 (1979): 503–6.

Taylor, D. D. "Differential Rates of Cerebral Maturation between Sexes and between Hemispheres." *Lancet* 2 (1969): 140–42.

Victor, M., R. D. Adams and G. H. Cullins. "The Wernicke-Korsakoff Syndrome." Philadelphia: F A Davis, 1971.

Waber, D. P. "Environmental Influences on Brain and Behavior," in *Sex Differences in Dyslexia,* eds. A. Ansara, N. Geschwind, and A. Galaburda, et al. Townsen, Md: Orton Dyslexia Society, 1981.